MATHEMATICS CURRICULUM, TEACHER PROFESSIONALISM, AND SUPPORTING POLICIES in Korea and the United States

SUMMARY OF A WORKSHOP

Ana Ferreras, Cathy Kessel, and Myong-Hi Kim, Rapporteurs

U.S. National Commission on Mathematics Instruction

Board on International Scientific Organizations

Policy and Global Affairs

The National Academies of
SCIENCES · ENGINEERING · MEDICINE

THE NATIONAL ACADEMIES PRESS
Washington, D.C.
www.nap.edu

THE NATIONAL ACADEMIES PRESS 500 Fifth Street, NW Washington, DC 20001

This activity was supported by Grant No. DRL1214813 from the National Science Foundation. Any opinions, findings, conclusions, or recommendations expressed in this publication are those of the authors and do not necessarily reflect the views of the organization or agencies that provided support for the project.

International Standard Book Number-13: 978-0-309-37436-1
International Standard Book Number-10:0-309-37436-7

Additional copies of this report are available for sale from the National Academies Press, 500 Fifth Street, NW, Keck 360, Washington, DC 20001; (800) 624-6242 or (202) 334-3313; *http://www.nap.edu.*

Printed in the United States of America.

Suggested citation: National Academies of Sciences, Engineering, and Medicine. 2015. *Mathematics Curriculum, Teacher Professionalism, and Supporting Policies in Korea and the United States. Summary of a Workshop*. Washington, DC: National Academies Press.

The National Academies of

SCIENCES · ENGINEERING · MEDICINE

The **National Academy of Sciences** was established in 1863 by an Act of Congress, signed by President Lincoln, as a private, nongovernmental institution to advise the nation on issues related to science and technology. Members are elected by their peers for outstanding contributions to research. Dr. Ralph J. Cicerone is president.

The **National Academy of Engineering** was established in 1964 under the charter of the National Academy of Sciences to bring the practices of engineering to advising the nation. Members are elected by their peers for extraordinary contributions to engineering. Dr. C. D. Mote, Jr., is president.

The **National Academy of Medicine** (formerly the Institute of Medicine) was established in 1970 under the charter of the National Academy of Sciences to advise the nation on medical and health issues. Members are elected by their peers for distinguished contributions to medicine and health. Dr. Victor J. Dzau is president.

The three Academies work together as the **National Academies of Sciences, Engineering, and Medicine** to provide independent, objective analysis and advice to the nation and conduct other activities to solve complex problems and inform public policy decisions. The Academies also encourage education and research, recognize outstanding contributions to knowledge, and increase public understanding in matters of science, engineering, and medicine.

Learn more about the National Academies of Sciences, Engineering, and Medicine at **www.national-academies.org.**

WORKSHOP PLANNING COMMITTEE

KOREA

Oh Nam Kwon, Cochair
Seoul National University

Mi-Kyung Ju
Hanyang University

Hyunyong Shin
Seoul National University

UNITED STATES

Patrick Scott, Cochair
New Mexico Inquiry Science Education

Gail F. Burrill
Michigan State University

Janine T. Remillard
University of Pennsylvania

U.S. NATIONAL COMMISSION ON MATHEMATICS INSTRUCTION

2014 ROSTER
(as of December 31, 2014)

Staff

Ana M. Ferreras
Senior Program Officer
Board on International Scientific Organizations

Pamela Gamble
Administrative Assistant
Board on International Scientific Organizations

Preface and Acknowledgments

In 2012, the 12th International Congress on Mathematical Education was held in Korea. The United States National Commission on Mathematics Instruction (USNC/MI) took advantage of this opportunity to organize a joint Korea–U.S. workshop on mathematics education, which was held at Seoul National University after the congress. It brought together about 40 teachers, mathematics educators, mathematicians, education researchers, and other mathematics education specialists from Korea and the United States.

During the workshop, speakers gave presentations on different aspects of education in the two countries: education policy; the interpretation of standards and curriculum guidelines in textbooks, teacher's manuals, and instruction; and the organization of teaching. A substantial portion of the workshop was devoted to discussion (see workshop agenda in Appendix A). Participants from both countries commented on aspects of the presentations that were notable from their perspectives as teacher, educator, researcher, or mathematician. They described details of their experiences and mentioned differences in educational systems that were new to planning committee members. This summary is an edited version of the workshop presentations and comments.

This is the third report of a bilateral USNC/MI workshop. The first two reports are *Studying Classroom Teaching as a Medium for Professional Development: Proceedings of a U.S.–Japan Workshop* (2002) and *The Teacher Development Continuum in the United States and China: Summary of a Workshop* (2010). All three reports are available for download, free of charge, from the National Academy Press website, www.nap.edu.

The workshop was planned and organized by a binational committee, with the assistance of a staff member from the Board on International Scientific Organizations (Ana Ferreras).

Support from the National Science Foundation and Seoul National University is gratefully acknowledged. Special thanks to Professor Oh Nam Kwon, Teacher Seoungyee Choi, and Professor Kwon's 10 graduate students—Hyungmi Cho, Mihye Cho, Ji Eun Lee, Eun Ji Kim, Young Hye Kim, Youngki Kim, Yujung Kim, Miyeong Na, Hye Mi Oh, and Kukhwan Oh. Professor Kwon not only supported the workshop from her grant but also hosted dinners that allowed informal discussion among participants. During the two weeks before the workshop, her students stayed up late to explore questions and locate information of possible interest to U.S. participants. During the workshop, they provided assistance that allowed it to flow smoothly.

This summary has been prepared by the workshop rapporteurs as a factual summary of the main presentations and discussion at the July 2012 workshop. The translations were made by Myong-Hi Kim. The statements made in this summary are those of the rapporteurs or individual workshop participants and do not necessarily represent the views of all workshop participants, the USNC/MI, or the Academies. Special thanks to Rita Johnson for editing and organizing the workshop participants' comments and preparing the report for release.

Acknowledgment of Reviewers

This report has been reviewed in draft form by individuals chosen for their diverse perspectives and technical expertise, in accordance with procedures approved by the National Academies of Sciences, Engineering, and Medicine. The purpose of this independent review is to provide candid and critical comments that will assist the institution in making its published report as sound as possible and to ensure that the report meets institutional standards for quality and objectivity. The review comments and draft manuscript remain confidential to protect the integrity of the process.

We wish to thank the following individuals for their review of this report: Roger Howe, Yale University; Ok-Kyeong Kim, Western Michigan University; Saeja Kim, University of Massachusetts; Catherine Lewis, Mills College; W. James Lewis, University of Nebraska-Lincoln; and Sunsook Noh, Ewha Womans University.

Although the reviewers listed above have provided many constructive comments and suggestions, they were not asked to endorse the content of the report, nor did they see the final draft before its release. The review of this report was overseen by Francis Fennell, McDaniel College; appointed by the Academies, he was responsible for making certain that an independent examination of this report was carried out in accordance with institutional procedures and that all review comments were carefully considered. Responsibility for the final content of this report rests entirely with the rapporteurs and the institution.

Contents

1
INTRODUCTION

On July 15–17, 2012, following the 12th International Congress on Mathematical Education in Seoul, Korea, the United States National Commission on Mathematics Instruction and Seoul National University held a joint Korea–U.S. workshop on Mathematics Teaching and Curriculum. The workshop brought together 40 teachers, mathematics educators, mathematicians, education researchers, and other mathematics education specialists from both countries.

Planned by a joint ad hoc committee, the workshop was organized to address questions and issues related to math teaching and curriculum that were generated by each country, including the following: What are the main concerns (or main focus) in the development of the curriculum? What issues have been discussed or debated among curriculum developers, teachers, teacher educators, and scholars regarding the curriculum? How have textbooks been developed for the curriculum? How are curricular tasks designed and what criteria are used? What is the role of learning trajectories in the development of curriculum? This volume summarizes the presentations and discussions at the workshop.

To understand differences in the educational systems of the Republic of Korea[1] and the United States, some general information about these nations is useful. According to the World Bank, in 2013 the population of the United States was approximately 316 million, more than six times that of Korea, which had a population of 50 million.[2] Although Korea's population is roughly twice that of Texas, its land area is much smaller—approximately the size of Kentucky or Virginia. The United States spends $10,995 per K–12 student in public education, while Korea spends $6,723, but outpaces U.S. academic performance.[3]

Korea has gained attention in international assessments of mathematical knowledge. The Trends in International Mathematics and Science Study (TIMSS) is administered to students in grades 4, 8, and 12. Tests and surveys for the Programme for International Student Assessment (PISA) are administered to 15-year-olds. Korea has been a top performer on these assessments for at least a decade. In contrast, the overall U.S. performance has been at or below the international average, although some U.S. states are among the high performers. On the 2012 PISA, for example, the states of Massachusetts and Connecticut had 19 percent and 16 percent of student scores classified as proficiency level 5 or greater. These percentages were above the international average of 13 percent,

---

[1] The Republic of Korea is the official name of "South Korea." Henceforth, this report refers to it as "Korea."
[2] The World Bank: 2014 World Development Indicators. Available at http://wdi.worldbank.org/table/2.1.
[3] See http://www.facethefactsusa.org/facts/money-cant-buy-genius.

and above the U.S. average of 9 percent. The corresponding figure for Korea was 31 percent. A similar pattern occurred for average PISA scores: 514 (Massachusetts), 504 (Connecticut), 481 (United States), and 554 (Korea).[4] On the 2011 TIMSS for grade 8, average scores were 561 (Massachusetts), 518 (Connecticut), 509 (United States), and 605 (Korea). The international average was 500. [5]

Although Korean performance on international assessments has received much attention in the United States, the structure of Korea's education system is far less well known. The accounts in this report offer an introduction to this system. They also provide an opportunity to gain a broader and more nuanced understanding of differences in policy, curriculum, and the profession of teaching, and their relationships in the Korean and U.S. education systems.

This wealth of information is grounded in the workshop participants' different experiences and types of expertise. Some of the Koreans had taught in U.S. schools, worked at U.S. universities, or were familiar with U.S. textbooks, and, in one case, had analyzed them together with one of the U.S. participants. Others, both researchers and teachers, were among the creators of national mathematics curricula, or had worked to embody these national guidelines in instructional materials or professional development offerings (see biographies in Appendix B of this report). Their experiences were reflected in their presentations and in detailed and thoughtful responses to questions from U.S. participants and comments of other Koreans. Among the U.S. participants were a teacher who had taught in two very different school districts; a teacher who now works in a state department of education; researchers who specialize in relationships between educational policy and classroom practice; and mathematicians with long-time interests in elementary mathematics.

This report summarizes a two-day workshop and should not be considered a complete account of the Korean and U.S. educational systems. Because the subject of the workshop was school systems, other educational institutions such as *hagwons* (private after-school academies) in Korea or test-preparation schools in the United States are not discussed in this report. This is not meant to imply that they may not wield considerable influence on student achievement.[6] Likewise, possible effects of college entrance examinations in both countries were raised during discussion, but were not the topic of a presentation.

Education Policy

One fundamental difference between the Korean and U.S. education systems is the distribution of authority. In the United States, many matters related to education are

---

[4] See *PISA 2012 Results in Focus* and http://nces.ed.gov/surveys/pisa/pisa2012/pisa2012highlights_1.asp.

[5] See http://nces.ed.gov/timss/results11_states11.asp.

[6] Four of five primary school children receive private education, typically in cram schools known as *hagwons*. Total private education expenditure for K–12 students in Korea was 20,100 billion won (approximately $20 billion) in 2011.
See http://monitor.icef.com/2014/01/high-performance-high-pressure-in-south-koreas-education-system and http://kostat.go.kr/portal/english/news/1/8/index.board?bmode=read&aSeq=254474.

under the jurisdiction of the states rather than that of the federal government.[7] There are 50 states and each has its own department of education.

Korea's educational administration system is centralized. Its offices are the central government's Ministry of Education, Science, and Technology (MEST), 16 provincial education offices, and 182 county and city offices of education. About 75 percent of the budget for local education offices is provided by the central government, which also underwrites private school deficits (Kim et al., 2008, p. 43). Public and private school teachers have comparable salaries and pensions (Kim, 2009, pp. 31, 42–43). Consistent with this, a workshop presenter noted that schools across the country are similar in teacher quality, facilities, and other resources.

Since the local autonomy law of 1991, local educational autonomy has been promoted. MEST has delegated much of its budget planning, personnel management, and administrative decisions to local authorities.

MEST funds several educational research institutes. Among them is the Korea Institute for Curriculum and Evaluation (KICE), which conducts basic and policy research on teaching and learning. KICE also develops and disseminates methods and materials for primary and secondary education. These include the assessments administered to all students in grades 6, 9, and 11, and college entrance examinations. Together with ordinary citizens, experts from schools, universities, and institutes like KICE participate in revising the national curricula (Huh, 2007, pp. 59–61).

Chapter 2 focuses on changes in the Korean national curriculum—in particular, the new goals of creativity and character building—and how these changes have been implemented in mathematics textbooks, teacher's manuals, and instructional guidelines. In Korea, the national curriculum framework for all subjects is developed first. Subject curricula are then written in alignment with the framework. These include objectives and content for each grade and guidelines on instruction and evaluation. Curriculum revision is a "complicated, time-consuming process" involving a multiyear plan for textbook revision and the provision of manuals and training for teachers (Huh, 2007, pp. 56–58). Implementation is left to the joint efforts of schools and provincial and local offices of education.

Since 1953, Korea has had seven national curricula.[8] In the past, curriculum changes tended to be wholesale, but since the seventh curriculum, changes have been incremental.[9] This history and the framework–subject curriculum structure are reflected in nomenclature. The Korean workshop participants referred to the current mathematics curriculum as "the 2011 revised mathematics curriculum" (referencing the year of its

---

[7] The Tenth Amendment of the U.S. Constitution states, "The powers not delegated to the United States by the Constitution, nor prohibited by it to the states, are reserved to the states respectively, or to the people."
[8] English translations of recent curricula are at the National Curriculum Information Center, http://www.ncic.re.kr/english.index.do.
[9] See *Education in Korea*, p. 6.

release) or "the 2009 revision" (referencing the year of the associated curriculum framework). See Table 1-1.

TABLE 1-1 Release Dates for National Curricula in Korea, as of July 2012

National Framework	Mathematics
First, 1954	
	1955
Second, 1963	1963
Third, 1973	1973
Fourth, 1981	
	1982
Fifth, 1987	
	1989
Sixth, 1992	
	1995
Seventh, 1997	
	2000
2007 revision	2007
2009 revision	
	2011

SOURCE: Bae et al., 2011; pp. 128–129; Park, 1997.

Chapter 3 discusses curriculum changes in the United States. In the United States, determination of curriculum guidelines and textbook criteria is distributed between state governments and school districts. Each state has a department of education, which sets performance standards. In many states, public schools receive a substantial part of their funding from taxes on property within their district. This variability in funding is part of the nationwide variability in schools and teachers.[10] Often, but not always, states, rather than districts, also set curriculum guidelines and textbook adoption criteria. Textbooks are created, for the most part, by commercial publishers. Development of research-based textbooks and other curriculum materials for mathematics is typically supported by private funders or the National Science Foundation. Timelines for initial development depend on funding guidelines (Schoenfeld, 1999, p. 11) as do subsequent evaluation and revision (Clements, 2007, p. 59).

The U.S. federal government and its Department of Education (DoED) have more limited policy roles than in Korea. The DoED's Institute of Education Sciences conducts the National Assessment of Educational Progress, which periodically tests a sample of students in grades 4, 8, and 12. Since 2001, the No Child Left Behind Act (NCLB) has made state receipt of federal school funding contingent on annual testing of every public school student in grades 3 through 8 and high school. In general, these tests have been created separately for each state.

[10] See, for example, Honegger, 2010. Table 7 shows amounts of federal, state, and local funding for various districts. Table 2 gives total spending per pupil per state.

In 2010, the Common Core State Standards for Mathematics and for English Language Arts and Literacy in History/Social Studies, Science, and Technical Subjects were launched. By the end of 2010, 41 states and the District of Columbia adopted the Common Core State Standards.[11] The development of the Partnership for Assessment of Readiness for College and Careers and Smarter Balanced Assessment Consortium as consortial assessments was supported, through a competitive grant process, by the United States Department of Education. These consortia worked with educators from around the country to develop tests aligned with the Common Core State Standards. Presentations and comments from U.S. participants mentioned issues related to teacher training and assessment, and the development of textbooks and related teacher manuals as possible obstacles to successful implementation of the Common Core State Standards.

Over the course of the workshop, several U.S. participants commented on the different policy arrangements in the two countries and their effects. For example, one participant noted the effect of having politically appointed education officials:

> The fundamental contrast between the two countries is that Korea has a ministry of education . . . staffed by mathematics educators, teacher professionals, and others who are relatively stable in their positions. They don't change with each election, and they're professionals who devote themselves to educational problems.

A Korean presenter mentioned that textbooks were piloted in designated schools and discussion touched on this practice. This also occurs in the United States for textbooks developed by education researchers, and, to a lesser extent, by commercial publishers. From a U.S. standpoint, it is perhaps more notable that Korea piloted policy changes such as new methods of school and teacher evaluation in designated schools before adoption.[12] Also, the evaluation methods themselves are quite different from those in the United States.

Several of the U.S. participants took note of a presenter's remark that for Korean teachers the authority of the textbook seems to be like that of the Bible. In contrast, one participant said, "U.S. teachers . . . teach whatever they want to teach, with great latitude."

Development of Instructional Materials

Chapter 4 describes the development of instructional materials for mathematics in Korea, including the organization of content. In the United States, the Common Core State Standards drew on research-based *learning progressions*. A learning progression consists of:

---

[11] How the content and organization of the new mathematics standards differs from that of previous standards was not discussed at the workshop.

[12] This was only briefly mentioned during the workshop. Pilots for the teacher evaluation program began in 48 schools in 2005. By 2009, pilots had expanded to 1,570 schools (Kim, 2009, pp. 73–76).

- A set of mathematical goals;
- A clearly marked developmental path; and
- A coherent set of instructional tasks or activities.

A related Korean term is *GaeNyumdo* (개념도, roughly "concept map"). Part of the presentation on instructional materials described how *GaeNyumdo* are used in developing curricular materials and the instruction that accompanies them. This was followed by discussion that raised the question "Are teachers given the conceptual framework that led to the development of the tasks?" According to several of the teacher participants, teachers in Korea are familiar with the *GaeNyumdo* used by curriculum developers. This led to a discussion of teacher's manuals and how they might support teachers' understanding ("speaking to the teacher") versus providing a script for teachers to follow.

A U.S. mathematician was struck by who was involved in developing instructional tasks—a team of mathematicians, mathematics educators, and mathematics teachers—and said that he would love for that to be routine in the United States. "We are expecting children to understand the concepts behind the math, but there might be teachers who don't understand them," said a U.S. teacher. Textbooks and manuals that support teachers' understanding can allow "teachers to grow with kids."

Teachers, Professional Learning, and the Organization of Teaching

Research on teacher knowledge suggests the importance of retaining teachers while giving them opportunities to learn. The structure of the education system appears to be an important factor in developing teacher expertise and specialized mathematical knowledge for teaching. In some education systems, teachers can acquire and increase this knowledge over the course of their careers.[13]

In Korea, teacher attrition is very low. Prospective teachers face stiff competition for admission to teacher preparation programs and in employment tests. New public school teachers who pass employment tests are initially hired as "second-rank teachers" by provincial offices, then complete "qualification training" before teaching. After further training and satisfactory evaluations, they can be promoted to first-rank teacher, master teacher, school inspector, research officer, or principal (Kim et al., 2008). Master teachers have a 40 percent reduction in teaching time, allowing them to help other teachers or carry out innovations in schools. Public school teachers rotate within their districts approximately every five years. Because many areas of Korea are densely populated, teachers often need not travel far to reach a different school.

[13] Ma (1999) compared samples of U.S. and Chinese elementary teachers, finding that many more Chinese teachers had specialized mathematical knowledge for teaching. More profound forms of this knowledge occurred among Chinese teachers with an average of 18 years of teaching experience, but not among the experienced U.S. teachers in the sample. Specialized teacher knowledge of mathematics is hypothesized as a necessary condition (Ma, 1999, 2013).

A workshop presenter mentioned that Korean teachers have a strong desire to learn and to increase their capacity as teaching professionals, describing how this desire is supported by MEST policies and MEST-funded activities. MEST funding is distributed by district offices to teacher research groups for the development of curriculum materials or innovative instructional methods and materials. Teacher research groups also provide professional development for other teachers.[14]

The Korean Society of Teachers of Mathematics (KSTM) is the first and largest teacher-created organization in Korea. It began in 1994 and is now one of three teacher organizations that are government-approved professional development providers. KSTM publishes a journal and also runs common interest groups, including a group for "lesson analysis." Chapter 5 describes the organization of teacher preparation and professional development in Korea, including an educational researcher's description of government-supported teacher research groups and mandated activities, followed by the KSTM founder's description of teacher-initiated activities.

Workshop discussion touched on a new strategy employed during the 2009 curriculum revision. MEST gave approximately $5,000 to 1,000 groups of 5 teachers to develop the national subject curricula. Following the curriculum revision, these groups were engaged in teacher training.

In the United States, there is a wide range of teacher preparation programs. Among sources of variance are selectivity in admissions and program requirements (see literature review and findings of Schmidt, Houang, and Cogan, 2012). A well-publicized statistic is "half of beginning teachers leave teaching within five years." This was publicized by the 2003 report *No Dream Denied*, which documented considerable turnover among U.S. teachers. In the workshop, this turnover was illustrated by a presentation from a teacher who was ready to quit teaching, but "fell in love with teaching again" when she moved to another district in another state. Her experience also illustrates research findings that suggest that school organization is a major factor in retaining teachers.[15] School organization is also a factor in supporting the existence of professional learning communities, an infrastructure or a way of working together that results in continuous school improvement (Hord, 1997). In addition to the account of a teacher who almost left teaching, Chapter 6 gives an overview of the landscape of professional learning communities in the United States.

In discussing these presentations, a U.S. participant noted that professional learning communities have existed for decades, but in isolated pockets. "We can create pockets of things done well, but it's very hard to do it at scale," said another. "There is no system for helping teachers, principals, and schools improve."

---

[14] Research groups and individual teachers may be recognized as winners of competitions sponsored by offices of education. Teaching competitions are reported on national television news. Three-time winners receive the title of "treasured teacher."

[15] Information about teacher preparation programs is summarized in the 2010 National Research Council report *Preparing Teachers: Building Evidence for Sound Policy*. Studies of attrition and turnover have been conducted by Richard Ingersoll and his colleagues. See, for example, the summary in Chapter 2 of *The Mathematical Education of Teachers II*.

One participant returned to the theme that U.S. classrooms tend to be places of isolation for teachers, saying "there's a history of teachers closing their doors, closing the world out, and doing things on their own." Another commented: "Now, the Common Core provides us an avenue to open up those doors." In contrast, one of the Korean teachers said, "We have quite well-established teacher communities. . . . Even with our busy schedules, we try to get together in our school or with nearby schools."

The discussion also touched on the consequences of student test results. In the United States, NCLB is "the continuation of a steady trend toward greater test-based accountability that has been going on for decades," according to the National Research Council report *Incentives and Test-Based Accountability in Education* (2011). A current example is "widespread interest in using student test scores as a way of rating and rewarding teachers and principals," which has been manifested in discussions of "value-added" measures.

Workshop discussion mentioned differences in Korean and U.S. systems for evaluation of teachers and methods of improving teacher quality. In 2010, after 10 years of piloting, the Korean government began using a new system for evaluating public school teachers. Findings from evaluations are used to create customized teacher training (Kim, 2009, pp. 73–76). Some details of evaluation and accountability were mentioned by Korean workshop participants. Schools keep a detailed, year-long log of educational activities. Monitoring committees who visit each school examine this log carefully, together with the principal. If a school is judged as providing a good quality education, it gets some extra funding. If a school has many low-achieving students, then the government provides support in the form of assistant teachers. Teachers themselves help to monitor teaching quality via observation and feedback on each other's lessons as required by regulations.

At the workshop, the comments of Korean teachers reflected a culture of teaching professionalism, in which teachers initiate their own study groups, conduct professional development, produce research and teaching materials, and collaborate with mathematicians and education researchers to produce curricular tasks. This suggests a type of teacher professionalism that seems rare in the United States. A Korean teacher said the workshop caused her to reflect on why she and teachers engage in these activities so much, wondering if it was due to cultural expectations for teachers, rather than a desire for self-improvement.

Like this teacher, readers will bring their own knowledge and experience to the accounts in the report, finding other aspects of interest. As often occurs, contrasts between nations provide a mirror, not a blueprint for improvement. They are an opportunity for reflecting on why each country does things the way it does, provoking closer examination of policy and practice.

2
MATHEMATICS CURRICULUM IN KOREA

One of the differences between the United States and Korea is the way in which curriculum change is initiated and implemented. For example, the Korean national curriculum framework describes goals to be achieved in each subject curriculum. "Fostering the creative human," a core objective of the current framework, has two components: character building and the development of creativity. Thus, the creators of the mathematics curriculum faced the question of how to interpret creativity and character building in mathematics instruction. Workshop presenters described how textbooks and teacher's guides have been designed to support the achievement of these and other goals specific to the mathematics curriculum, in particular, learning of fractions. These presentations were followed by a reflection from Do Han Kim, president of the Korean Mathematical Society, giving his perspective on the role of mathematical societies and the problems of Korean education.

After the presentations and over the course of the workshop, U.S. participants commented on aspects that were notable to them. Among other things, they mentioned achievement gaps and pilot schools, and the idea of creativity for all students. They asked questions about the meaning of character building (인성교육), leading to more extensive discussion. An edited version of this discussion appears at the end of this chapter.

Textbooks as Mediators of the National Curriculum

Oh Nam Kwon of Seoul National University discussed how a new subject curriculum is created, how it is related to the national curriculum framework, and how it influences what happens in the classroom.

In Korea, the authority of the textbook seems to be quite different than in the United States. It can almost be compared to the Bible, said Kwon. Textbooks and teacher's guides are an important mediator between the national mathematics curriculum and what happens in the classroom. Instruction is textbook based rather than curriculum based. Because of this, it is important to develop high-quality textbooks and curriculum materials that are tightly coupled with changes in the national curriculum.

The national curriculum framework describes overarching goals and grade-level structure, and gives instructions for implementation. These are elaborated in the mathematics curriculum, which gives detailed specifications for:

- Learning goals by school level;
- Grade-level content objectives;
- Benchmarks;

- Instructional methods; and
- Assessment.

Guidelines for writing textbooks, teacher's manuals, and student workbooks are then written to be consistent with these specifications.

The 2011 Mathematics Curriculum Revision

Hee-Chan Lew of the Korea National University of Education discussed events that informed the revision of the mathematics curriculum and key accompanying policies.

The current curriculum focuses on creativity and personality. Society requires that students become not only more creative and competent but also more rational and sensible, so that they can comply with the rules and orders of the greater society. The previous curriculum included competencies that are believed to be fostered by engaging in mathematical processes such as mathematical reasoning, problem solving, and communication.

But, Lew said, speaking as a writer of the national mathematics curriculum, we have recognized another important factor in mathematics education: mathematical attitude. Korea has performed well on the two international assessments: the Trends in International Mathematics and Science Study (TIMSS) and the Programme for International Student Assessment (PISA). On the mathematics PISA, for example, Korea was ranked second in 2003, third in 2006, and fourth in 2009. But Korean students' attitudes toward mathematics were among the worst in the world, according to surveys that accompanied the assessments. Moreover, analysis of scores for the 2003 and 2006 PISA showed an achievement gap according to school location: Urban students scored much higher than rural students. In urban areas, Korea was among the best in the world. But the more rural the area, the lower the scores; in villages and small towns, achievement was below the international average for PISA countries.[1]

Why so different? Korea is a small country. Teachers, curriculum, textbooks, and school facilities are almost the same across the country. Lew believes that differences in educational outcomes may come from differences in mathematical attitudes of students and parents.[2]

For Lew, these findings are part of the background for the curriculum revision. In his view, key points of the revision and key accompanying policies are:

- Contextual learning from which students grasp concepts and make connections to everyday life;
- Manipulative activities through which students may attain an intuitive idea of what they are learning and enhance their creativity;
- Reasoning to justify mathematical results based on students' knowledge and experience;

[1] Much of the Korean population is concentrated in metropolitan areas, with about 20 percent in Seoul. Lower-income Koreans tend to live in rural areas rather than cities.

[2] There is a socioeconomic gap between urban and rural regions. Pahlke, Hyde, and Mertz analyzed Korean grade 8 results from TIMSS 2003 and 2007. They found that the father's education and family possessions, as well as the economic disadvantage of the school, were all significant predictors of mathematics and science performance.

- Reform of textbooks and the classroom environment: emphasis on “storytelling” as a strategy for teaching science, technology, engineering, and mathematics; and
- Emphasis on the professionalism of mathematics teachers, for example, sending 1,000 teachers to the 12th International Congress on Mathematical Education.

In the mathematics curriculum, content has been decreased and more emphasis has been placed on modeling.[3] Lew believes that the reduction will significantly ease students’ study load, and give more time for creative activities and fostering positive attitudes toward mathematics.

The goal of this reform is that, with only few exceptions, all students should be able to gain the mathematical confidence and competence required for private and public activities, and the mathematical skills necessary for their careers. They should also recognize the social and cultural importance of mathematics.

Mathematics Curriculum Design

JeoungSuk Pang of the Korea National University of Education described important aspects of recent curriculum frameworks reflected in the design of mathematics curricula. Pang and other workshop attendees participated personally in the national curriculum revisions. She identified four key aspects:

- Objectives (why to teach/learn)
- Content (what to teach/learn)
- Progression (when to teach/learn)
- Instruction (how to teach/learn)

The basic objectives that have remained in national mathematics curricula are the following:

- Acquisition of mathematical knowledge and skills
- Enhancement of mathematical thinking ability
- Cultivation of problem-solving ability and attitude

In the 2007 revision of the mathematics curriculum, more objectives were added:

- Mathematical communication
- Positive attitude

Mathematical communication was included because it may deepen mathematical thinking and because Korean students are perceived as passive and silent. Although positive attitude is affective and the curriculum places more importance on cognitive abilities than affective aspects, Pang said, these could not be ignored after the release of the TIMSS and PISA findings about Korean students’ attitudes toward mathematics.

---

[3] More details about the reduction are in Lew et al., 2012.

New objectives of the 2011 revision were as follows:

- Mathematical creativity
- Character building

Mathematical creativity is especially emphasized for national competitiveness. In the Korean mathematics curriculum, it is construed as having three subabilities: (1) problem solving, (2) communicating, and (3) reasoning. Because of the characteristics of mathematics as a discipline, character building was not previously emphasized in mathematics instruction. But school violence has been a serious social issue recently, and it could no longer be ignored.[4]

Pang said with respect to content, since the fourth curriculum (1981), the general tendency in official curriculum documents has been to reduce it. But in practice, curriculum designers have only shifted content across grade levels. For example, in the seventh curriculum (1997), the intent was to reduce the content by 30 percent, but mathematics curriculum designers did not comply with this policy. Optional topics for advanced students and material intended for underachieving students were taught to all students because there is a strong belief that all content in the textbook should be covered in mathematics lessons. The result is a heavy study load for students and a high teaching load for teachers. Since 2012, this tendency has more serious consequences because the number of schooldays per week has been reduced from five and a half to five. Without any change in the curriculum, this would leave no room for mathematical processes.

Pang said that the writers of the 2011 mathematics curriculum tried to delete optional topics and topics that were mentioned only once or isolated from others. The remaining topics were reorganized in five grade bands: elementary (1–2, 3–4, 5–6); middle (7–9), and high school (10–12). For instance, numbers were dealt with more effectively. Four-digit numbers, for example, were put in the first grade band, rather than being distributed over three grades: numbers up to 100 in grade 1, up to 1,000 in grade 2, and up to 10,000 in grade 3. So, said Pang, they think of teaching fewer topics in greater depth, instead of teaching more topics in a cursory manner.

With regard to progression—when to teach—greater flexibility as to when specific mathematical topics are to be taught has been gradually introduced. In 1997, the mathematics curriculum specified grade levels and even the semester level. Similar topics were separated inefficiently. In 2011, however, schools were given greater flexibility in organizing topics within grade bands.

Concerning when to teach, two other aspects were considered: (1) the difficulty of a given topic and (2) relationships among related topics. For example, relationships among quadrilaterals were moved from grade 4 to middle school. Formerly, the different concepts of fractions were spread over grades 2 to 5. That range will contract to the second grade band (grades 3 and 4).

[4] School violence in Korea includes bullying and fistfights, but not guns (which are illegal in Korea). See the Ministry of Education, Science, and Technology's *2014 Report on School Violence: Survey and Analysis*.

In mathematics curriculum documents, four instructional objectives have been consistently emphasized:

1. Meaningful questions
2. Mathematical concepts and principles
3. Problem-solving ability
4. Students' positive disposition

Comments were added to the 2011 mathematics curriculum about methods to enhance communication, thinking and reasoning abilities, and students' creativity and character building.

Pang gave more detailed explanations of the four objectives from the 2011 mathematics curriculum:

1. Students acknowledge that communication is crucial in learning and using mathematics by their actions of clarifying and reflecting on their thinking through representing and discussing mathematics.
2. To enhance mathematical thinking and reasoning ability, students infer mathematical facts with induction or analogy, and justify or prove them.
3. To nurture mathematical creativity, mathematical tasks are used to produce different ideas, stimulating students' divergent thinking. (In the past, creativity has been associated mainly with gifted students, said Pang, but it is relevant to all students. Also, there is a difference between mathematical creativity and general creativity.)
4. To nurture character building, students respect different solution methods and opinions posed by their peers.

So, what are the emergent challenges? Pang listed five:

1. A thorough review of curriculum via systematic research, considering two questions: What are the essential topics? How can educators assess the efficacy of curricular changes?
2. How can teachers promote mathematical processes in their students?
3. Use of differentiated instruction is another challenge. Since the seventh national curriculum, Korea has tried to use differentiated instruction, but has met with much resistance from students and insufficient numbers of teachers for implementation.[5] So, the national curriculum left decisions about implementation to individual schools.
4. How can teachers change students' negative dispositions about mathematics, considering the learner's perspective and the prevailing pressure of the high-stakes college examination in Korea?[6]
5. Although important, an effective curriculum does not suffice if other elements of the educational system are not in place. The challenge is to connect an effective mathematics

[5] A workshop participant said that the attempt to use differentiated instruction was implemented by splitting one or two classes of students into smaller groups according to student level, each with one teacher. For example, two classes were temporarily split into three levels, each with its own classroom.

[6] According to the Korea Institute for Curriculum and Evaluation's *Education in Korea*, "Since entering a good university is a matter of high importance in Korea, the evaluation content and method of CSAT [College Scholastic Ability Test] have great influence on the content and method of education in elementary, middle and high schools."

curriculum to high-quality instructional materials, efficient classroom teaching by well-educated teachers, and students' performance.

Creating Elementary Textbooks for the 2011 Curriculum

Man Goo Park of the Seoul National University of Education described how the 2011 revision of the mathematics curriculum has been reflected in elementary textbooks. His institution specializes in the education of preservice elementary teachers, and he is an experienced member of an elementary textbook writing team. He noted that, so far, Korea has had only one elementary mathematics textbook series, but the Ministry of Education, Science, and Technology (MEST) has changed its textbook adoption policy to allow the existence of more than one elementary textbook series.

Under the new policy, MEST provides stronger guidelines for textbook writers, including the number and size of pages. This constrains the approaches that can be used in textbooks. Mathematics textbook writers have to think about how to cover the content of the curriculum within those constraints.

There was concern that some textbooks were not being tested in real classrooms. The 12 experimental schools in which textbook material has been piloted before publication generated many problems with the textbooks for the writers to solve, but it is not easy to revise textbooks within the time lines.

Another constraint for textbook writers is how to implement the new instructional approaches, like storytelling, in mathematics textbooks. The spirit of introducing storytelling is wonderful, said Park. But it is not easy to integrate mathematical concepts into storytelling.

Another concern is how detailed the textbook content should be. Is it better to have less or more? Similarly, there needs to be a balance between classroom centeredness and student centeredness.

Park showed an example from his first-grade textbook, which is still under development. The section presented was about the concept of rhombus, used real-world situations, and has been much revised. The textbook illustrates a carjack lifting a very heavy car and a rack for drying clothes. Students are asked to think about the rhombuses within the carjack and clothes rack. At the end, there is an application section asking students to think about designing clothing or jewelry. (Park showed an illustration of a sweater, ring, and chain-link fence, all with a rhombus pattern.) It asks students to look around them, find examples of tools that use rhombus shapes, and talk about the features of those tools and how they are used.

Reflecting Curriculum Goals in Instruction

Rae Young Kim of Ewha Womans University spoke about how curriculum objectives have changed since the 1980s, how those changes have been reflected in instructional materials and methods, and issues arising from the process.

Curriculum trends in the United States and Korea were similar until the 1980s.[7] In 1992, as Korea became aware of international comparison studies, its sixth curriculum started to deviate from that of the United States. Kim focused on how changes in the four most recent Korean curricula (the sixth and seventh curricula, and 2007 and 2009 revisions) were reflected in textbooks.

Key changes in overarching curriculum goals have been more emphasis on problem solving and increased movement toward student-centered instruction. Recently, there has been a focus on strengthening mathematical communication.

Since the sixth curriculum, problem solving has been emphasized, and this emphasis has been reflected in the improvement of the instructional method and assessment. The seventh mathematics curriculum added an emphasis on strengthening mathematical reasoning and mathematics in daily life. It initiated emphasis on student-centered curriculum and suggested differentiated curriculum tailored to students' level of ability and lessons that fit students' ability. Consistency in instruction, textbooks, and assessment were emphasized.

Content changes included selection and arrangement of content as well as changes in approaches to the content. One change in approach has been increased emphasis on problem solving. For example, more tasks that ask students to create their own problems.

Other changes have been pedagogical. Since the seventh curriculum, there has been a movement toward student-centered learning. There has been a focus on individualized instruction (e.g., differentiation based on student level), and concern about motivation and inculcating positive attitudes toward mathematics. Other changes involve the use of technology in instruction and changes in assessments.

In the 2007 revision, the emphasis was on mathematical communication and positive attitudes. Before 2007, curricula described the same objectives for all grades. In 2007, the objectives differed by school level: elementary, middle, and high school. Attention to students' aptitude and their way of thinking was mentioned. At this point, the 2009 curriculum had been published, but textbooks for it had not yet been developed.

To move toward student-centered learning (a focus of 2007), emphasis on algorithms in elementary textbooks was shifted to let students choose computation methods, and to promote more variety in solution approaches and group collaboration (textbook instructions include "Let's do it together" and "Do it yourself"). Between the seventh curriculum and the 2007 revision, there was increased emphasis on communication via classroom discussion. To promote mathematical communication skills, the 2007 textbook does not stop with asking elementary students to solve problems but also asks for explanations of how they solved the problems, orally ("Try to explain") as well as summarizing in writing ("Try to write"). For example, a textbook

---

[7] According to *Education in Korea*, immediately after Korea's liberation from Japan in 1945, the United States Army Military Government in Korea stipulated education policy. Later curricula followed U.S. trends. For example, the main focus of the third mathematics curriculum (1973–1981) was "new math" and the fourth (1982–1988) was "back to basics" (Pang, 2014; Park, 1997).

problem says: "Here is the work of Young and A-Ram for the same problem. Try to explain their methods." It shows (–2/3 × 1/4) ÷ –5/6, followed by two different computations.

Changes in Content

Jee Hyun Park, a teacher at Seoul Finance High School, described issues of selection and arrangement of content raised by curriculum changes. She gave two examples.

The first example concerned vertical consistency in the case of fractions. As noted earlier, in the 2011 curriculum, concepts of fractions contracted to the second grade band (grades 3–4). Formerly, these were distributed and introduced between grades 2 and 6:

- Grade 2, division of a single-unit whole.
- Grade 3, parts of a multiunit whole.
- Grade 4, proper, improper, and mixed fractions.
- Grade 5, addition, subtraction, and multiplication of fractions.
- Grade 6, division of fractions.

In later curricula, content was reduced or removed to optimize learning. For example, between the seventh curriculum and the 2007 revision, the binary system and correlation were removed, proofs that involved the application of similarity were reduced, and direct and inverse proportion was shifted to upper grades.

A second example of change was in conceptual focus—how concepts were construed. For example, in the sixth curriculum, functions were interpreted as correspondences between two variables, in the seventh they were approached via proportional relations, and in the 2007 revision, as relations among variables.

Changes in Pedagogy

Jung Sook Park, a teacher at Taenung High School, described pedagogical changes made in response to national curriculum changes.

To illustrate changes for individualized instruction, Park contrasted pages of a textbook for the seventh curriculum with a student workbook for the 2007 revision. The latter showed a row of three colored dots above each problem to indicate the difficulty level of each problem and allow students to choose problems according to their abilities.

To promote positive attitudes toward mathematics, 2007 revision textbooks included cartoons to stimulate interest, used puzzles to practice terminology, employed approaches using empirical experiments, and inserted photos of students to give a sense of realism.

The use of technology changed. In the 2007 revision, computers were used as illustrative tools in instruction as well as in homework assignments. E-books were developed alongside paper textbooks.

Assessments changed. The seventh curriculum textbooks only give problems. The 2007 revision textbooks include performance assessments, asking students to make up their own problems or to analyze data. For example, after an illustration using a spreadsheet to manipulate data in a frequency table, a textbook directs, "Collect data from everyday life. Using a computer program, represent your data as a frequency table and in various graphical forms. Then analyze the data."

From the sixth curriculum on, the vision has been student centered, with the use of real-life contexts to try to encourage creative problem solving. The trend is emphasis on mathematical reasoning and classroom communication.

Textbook Development, Approval, and Implementation

Mi-Kyung Ju of Hanyang University described the process for textbook development, approval, and implementation. One focus in textbook development is mathematical personality traits, including creative problem solving and various forms of mathematical communication. A second focus is character building.

In 2009, the national curriculum framework was released, and in 2011 the mathematics curriculum was released. In 2012, Korea was developing textbooks for the mathematics curriculum. There are three types of textbook adoption processes:

- Type A. MEST is the author.
- Type B. A publisher develops the textbook and is its author. The textbook is approved by MEST.
- Type C. A publisher develops the textbook, and is its author. The district approves the textbook.

The trend has been that type C has increased, and types A and B have decreased. In the recent past, all elementary textbook adoptions were of type A. Middle and high school adoptions were of types B and C.

Table 2-1 describes the time line for textbook development, approval, and adoption. Textbooks were to be implemented incrementally, by grade band. Textbooks for the first grade bands at each level, that is, for grades 1–2 and 7–9 were to be developed first. The plan was for textbooks for these grade bands to be submitted for approval in August 2012, the results were to be announced in October 2012, and implementation was to occur in spring 2013. Table 2-1 shows the time lines for later grade bands. Textbooks for the last grade band will be implemented in classrooms in 2015.

TABLE 2-1 Time Lines for Textbook Development (D), Approval (A), and Implementation (I)

School level	Grade band	2011	2012	2013	2014	2015
Elementary	1–2	D	D/A	I		
	3–4		D	D/A	I	
	5–6			D	D/A	I
Middle school	1–3	D	D/A	I		
High school	1–3		D	D/A	I	

MEST formed a committee for textbook policy that solicited opinions from teachers, policy makers, and mathematics education experts. Three recommendations were prominent:

- More autonomy and variety needed to be allowed in textbook development.
- Textbooks should reflect academic and social shifts promptly.
- Textbooks should be accessible (i.e., "student-friendly," readable), interesting, and familiar to students. Material should be developed to promote students' interest in mathematics.

In response to the recommendations, all textbook adoptions for the 2009 curriculum were of type C. Specific guidelines were removed, such as textbook size, color, paper quality, and length. The approval criteria were changed to allow more autonomy and variety.

A system that allows selection and reorganization of content to meet current needs was recommended, so that textbooks reflect academic and social shifts promptly. Also, more teachers should participate as textbook authors, since they are the ones that can see needs and shifts in the classroom. Textbooks should be changed and approved on a flexible, rather than a fixed, schedule.

Ju concluded by listing authors' responses to the recommendation that textbooks should be easy, interesting, and familiar to students:

- Use of real-world problems to introduce and develop mathematical concepts and principles.
- Development of aligned materials for teaching and learning of mathematics.
- Development of mathematics textbooks based on storytelling, to provide students opportunities to be absorbed in doing mathematics via activities and projects.

Reflection on Korean Education

Dohan Kim, the president of the Korean Mathematical Society, made some brief remarks about Korean education.

He noted that when the seventh curriculum was introduced, both the mathematics and mathematics education communities were very shocked. The seventh curriculum benchmarked Japan's *Yutori* curriculum and reduced the content of school mathematics by one-third. Japan had lower student achievement with the *Yutori* curriculum, but soon acknowledged its failure. In Korea, however, the mathematical community as a whole was not aware of this failure for a long time.

In response to this event, the Korean Federation of Mathematical Communities formed to address policy as a united professional organization. The federation's members include the Korean Mathematical Society, Korean Society of Mathematical Education, Korean Society for Industrial and Applied Mathematics, and Korean Women in Mathematical Sciences.

Kim said that in the past, when there was a curriculum reform or other change, government representatives did not listen to mathematicians, saying "How can you, the representative of one organization, represent all communities?" But nowadays, the federation brings experts in the field to meet government representatives and they listen to the federation's suggestions most of the time.

Kim described what are, in his view, the problems of Korean education. It is very competitive. Parents let their children learn ahead of the school curriculum. Some teach their children calculus before they enter high school. Politicians use education to score political points, but opportunity for experts to express their opinions has lessened.

He hopes that experts in mathematics education express their opinions via professional organizations and journals, and that Korea improves education for advanced students as well as for those who fall behind.

Comments

Workshop participants offered a wide range of comments on a variety of topics, including the following:

Similarities between the countries

- Both countries are trying to reduce content and cover subjects more deeply.
- Both countries emphasize differentiation and try to meet the needs of individual students.
- Both countries are moving toward e-books, and are increasing the use of online textbooks.

Attitudes toward textbooks

- Korean teachers cover all the material in the textbooks, while in the United States, teachers act as independent contractors and teach whatever they want to teach with great latitude.
- MEST wants the textbook to be one of many teaching-learning materials, lessening its status.

Character building (Korea)

- In the national curriculum framework in Korea, the objective of "character building" had its origins in the school subject of ethics. The meaning of the English term "character building" includes honesty, keeping one's word, being considerate of others, and cooperating with others. Character building does not belong in the textbook, but in instruction and should be embedded in the curriculum. Ways of accomplishing this are still in the exploration and research stage.
- One example of embedding character building is the storytelling lesson structure that was developed, which is intended to stimulate imaginative ability. It is called "creating story structure." Students tell stories that fit given mathematical content. They then collect information from the outside world and make their own stories. Making a mathematical decision about whether data and information in the newspaper really speak the truth mathematically is one possible approach.
- In elementary school, there are two instructional methods for character building: One is indirect, involving why we do math, mathematics as a human activity, and the pursuit of precision and accuracy. The second method is to use cooperative work during mathematics instruction to illustrate diverse ways one can think and to ask students to suggest two or three methods, consider which shows better thinking and whether it can help them solve other problems.
- From an elementary school perspective, the context is selected to develop character, such as helping out in nursing homes, where teachers can talk about character naturally. The other is using properties of mathematics itself, such as the properties of the identities 1 and 0, such as: $x \times 1 = x$, $x \times 0 = 0$. In relation with others, there are personality types, such as self-absorbed or absorbed in others.
- Since teachers teach all subjects at elementary schools, a teacher can use interdisciplinary teaching, combining math and social sciences. But in secondary school, it is not easy because math teachers teach only math courses. It is impossible

to realize this goal unless a curriculum combining math and art, math and science, and so forth, is created and appropriately assessed.

- Giving Korean students some open exploration did not result in any noticeable drops in academic performance. Classroom teachers notice that students find math more interesting with this approach.
- Giving Korean students the opportunity to name figures, such as the example of the isosceles triangles is very nice, but that kind of work is advancing their creativity more than character building. Character building may be related to what people like Paul Cobb[8] and his colleagues call "sociomathematical norms"; that is, how do children behave when they do mathematics in a classroom? How is a student treated if he or she makes a mistake? That makes a great deal of difference in the student's feeling about participating in mathematics in the classroom. When you want to cultivate a culture of reasoning and proving—of justifying mathematical claims—you want to treat the classroom as a kind of intellectual community. Listening critically to the ideas of others. That is a mathematical process but it also very deeply involves human behavior and interaction. That is where you see character building integrated with mathematical work.

Student authorship (United States)

- In the United States, the closest thing to character building is "developing students' authorship or their own voice," finding out they can do mathematics and that it is a sensible pursuit. U.S. teachers have sometimes gone overboard in that direction, and everything and anything that students say is praised and carried forward.
- In the United States, it is challenging to maintain accountability to the discipline along with student authorship. It is called "the tension between authorship and accountability."

Teachers' reactions to curriculum revisions

- The time line for going from curriculum revision to development of materials and to implementation is very fast compared to that typical in the United States. Every time a new curriculum is implemented or changes are made in Korea, there is professional development among teachers and the district sends material helpful for teaching. Districts also provide lots of teacher training focused on how to assess via teacher activity, with references, and examples of assessment.
- In Korea too, there is a constant battle between the teacher-centered, or academic, approach and the student-friendly approach. Korea is strongly influenced by mathematicians. Another factor dominating Korea is the college entrance exam. Korean teachers ask themselves, "How can I help my students achieve higher scores in college exams in order to go to the best college?"
- The 2009 revision in Korea was called the "Creativity–Character Building Curriculum" (CCBC). Previously, the finished curriculum and curriculum materials were delivered to the districts. For the 2009 revision, the Korea Foundation for the Advancement of Science and Creativity recruited 1,000 groups, each consisting of four or five teachers, and provided support for these groups for materials

[8] Yackel, Erna, and Paul Cobb. 1996. "Sociomathematical Norms, Argumentation, and Autonomy in Mathematics." *Journal for Research in Mathematics Education* 27 (4):458–477.

development. This differs from the past practice of implementing an already-developed curriculum in a top-down fashion. This revision was middle-to-top and middle-to-bottom. The task force of teacher groups played a bridging role by providing professional development for other teachers. Principals, assistant principals, and school inspectors also attended. For elementary-level implementation, one of the 1,000 CCBC groups visited each school and provided professional development to all of the teachers in that school, not only in mathematics but also in other subjects: art, music, and so on. They explained how creativity and character building in mathematics can be interpreted and how they can be embedded in teaching. They provided various kinds of materials and explained the theoretical background. Teachers were interested and engaged in learning.

3
MATHEMATICS CURRICULUM IN THE UNITED STATES

The presentations described in this chapter begin with an overview of the education system in the United States and constraints on curriculum reform. In particular, local control of the curriculum requires that multiple influences be aligned if educational reform is to occur. This overview is followed by comments and discussion from all participants. Next are two reflections on the Korean and U.S. presentations. The first is a U.S. perspective from Hyman Bass, a past president of the International Commission on Mathematical Instruction. It describes how the current reform (the Common Core State Standards [CCSS, or Common Core]) occurred in a manner consistent with the principle of local control and gives an overview of the new standards. The second is a European perspective from Gabriele Kaiser, the convener of the 2016 International Congress on Mathematical Education (ICME). This notes the similarity of curricular traditions in continental Europe and Korea and how they differ from those in the United States. These reflections are followed by presentations on the implementation of the current reform and comments from the workshop participants.

Overview of Influences on Curriculum

Janine Remillard of the University of Pennsylvania and Brenda Gardunia of Frank Church High School, Boise, Idaho, gave an overview of the education system in the United States and identified what they considered key issues. To illustrate how the U.S. system operates, they discussed how schools are organized and how the curriculum is structured and described two efforts at curriculum reform.

Janine Remillard began with an outline of the presentation, noting that there are some significant differences in the way in which the Korean and U.S. educational systems are structured. The presentation was intended to give a sense of the structure of the U.S. system and in particular how decision making happens, where the authority lies, and why making changes in the United States education system is very challenging.

Remillard discussed four key features of the system: (1) an emphasis on local and regional control of the curriculum; (2) the limited role of the federal government; (3) the distributed, rather than centralized authority; and (4) the strong market influence of commercial publishers.

The United States comprises 50 states and several territories. Each state has a department of education and each state consists of one or, often, many school districts. These vary in size and are determined by region. In each state, the state department of education sets the standards that determine the goals for students and the assessments that are used to gauge whether students have met those goals. The local school districts can select and adopt textbooks to use and provide other instructional guidance. So, much decision making occurs locally.

The United States does not have a ministry of education. The Department of Education (DoED) is part of the federal government, but it has very limited control over educational decision making. The DoED tends to influence educational decisions in two ways. One is through funding. Less than 10 percent of funding at the state level comes from the federal government, but DoED can set conditions for receipt of funding. Second, DoED can influence policy by commissioning reports. These can raise concerns and identify issues that need to be addressed, but it is the responsibility of the states to take action.

Three major publishers dominate this market. These are Pearson, Houghton Mifflin Harcourt, and McGraw Hill, who, together, are responsible for around 90 percent of textbook sales in the market.[1] They produce a wide range of textbooks that local school districts can adopt and that influence the curriculum.

Also, there are a number of what Remillard called "nongovernmental organizations" with a stake in education: academic organizations; professional organizations, such as the National Council of Teachers of Mathematics (NCTM); and political organizations, such as the National Governors Association; and other sorts of organizations. These organizations tend to exert their influence by providing expert commentary, by undertaking research and presenting research findings, and sometimes by offering solutions to states and school districts.

Another influence is funding agencies. Some are federal, such as the National Science Foundation (NSF), and some are not. By providing funding opportunities, they can influence what happens in states and school districts by providing resources for curriculum development, as well as for professional development and other activities that happen at the local level. These agencies often work together with some of the nongovernmental organizations.

How Schools and Curriculum Are Organized

Brenda Gardunia gave an overview of how schools are organized and how the curriculum is organized. It is difficult to determine what is typical for the United States, so Gardunia described what she considered to be typical in a majority of the schools.

The United States has elementary, middle, and high schools. Typically, elementary schools start with kindergarten and progress through grades 1, 2, 3, 4, and end at grade 5 or 6, when children are 10 or 11 years old. In elementary schools, students stay in one classroom with one teacher who is certified to teach all the subjects that elementary students learn. At some schools, a specialist comes in to teach subjects such as art or physical education.

Middle schools typically include grades 6 through 8. Students are ages 12 to 14. They generally go to different classrooms and different teachers for each subject. Each teacher stays in one classroom. These teachers may have an elementary certification in all subjects or a secondary certification, usually for one subject area.

[1] Stephen Noonoo. How "Big Three" Publishers Are Approaching iPad Textbooks. *The Journal*, 2012.

Students attend high school from grades 9 through 12. Grade 9 is sometimes in the middle school building, and sometimes in the high school building. High school teachers are certified in their subject area. As in middle school, students move from classroom to classroom for different subjects. In high school, students accumulate the credits that they need toward their graduation requirements. And students move into different ability-based mathematics course sequences, which Gardunia referred to as basic, standard, and advanced.

To describe how the curriculum is typically organized, Gardunia began with the K–8 curriculum. It comprises five different content strands:

- Number and Operations
- Data and Probability
- Geometry
- Measurement
- Algebraic thinking

The curriculum is organized in spirals. Each strand is addressed at each grade as students go through elementary school. Ideally, each time students are exposed to a strand, they experience it in more depth and at a more sophisticated level. This organizational approach also includes process strands, such as problem solving, reasoning, and communication.

In high school, most schools offer Algebra I in grade 9, Geometry in grade 10, and Algebra II in grade 11.[2] Some schools offer an integrated sequence for grades 9 to 11. In both cases, mathematics is often optional in grade 12, and students have a choice of different courses such as Precalculus, Calculus, Statistics, or Discrete Mathematics.

Many schools permit advanced students to enroll in Algebra in grade 8, receiving high school credit for this course. This is followed by Geometry in grade 9, giving students the opportunity to take Calculus before graduating from high school.

Authority for determining graduation requirements varies. A state, district, or sometimes even a school may decide graduation requirements. Some require only two years of mathematics. Most require three years of mathematics, but an increasing number of schools require four years, said Gardunia.

Two Reforms

The next part of the presentation gave two examples of curriculum reform in the United States, illustrating the four key features mentioned at its beginning.

Remillard described the reform that occurred more than 20 years ago with the NCTM's *Curriculum and Evaluation Standards for School Mathematics.* These standards were developed

---

[2] The National Assessment of Educational Progress states that 47 percent of grade 8 students reported taking "advanced math." See http://www.brookings.edu/research/reports/2013/03/18-eighth-grade-math-loveless.

in response to a report that was produced in 1983 called *A Nation at Risk,* which was one of the federally commissioned reports mentioned earlier. It was intended to raise concern about the quality of education in the United States.

The NCTM responded to that report by putting forth a set of standards that called for many changes. In very brief summary, some of the key features were as follows: increased attention to conceptual understanding, problem solving, and reasoning; and decreased attention to the teaching of rote procedures.

The NCTM is an organization that is outside the government, a professional organization without the authority to set policy. However, in state departments of education, there were mathematics curriculum specialists who were members of the NCTM. When the NCTM produced its 1989 standards, state departments of education gradually embraced or slightly adapted those standards and put them in place in their states.

NSF was sympathetic to the calls for change by the NCTM standards and funded a number of projects to develop new curriculum materials that would help teachers instruct in alignment with the NCTM standards.

In those projects, curriculum developers spent several years producing materials. They also established relationships with commercial publishers who marketed the materials to local school districts. That is how support for and examples of what the NCTM standards might look like in a classroom found their way to teachers in local school districts. It is very different from a ministry of education saying: "This is the textbook."

NSF also funded professional development projects within states and local districts. These projects were intended to help teachers learn about the new standards and teach mathematics differently.

Gardunia described a second example of reform—the Common Core State Standards for Mathematics, which are different in that they are common standards across the United States. (In 2012, 45 states and three territories had formally adopted the CCSS.) The CCSS was a state-led effort coordinated by the National Governors Association (NGA) and the Council of Chief State School Officers, developed in collaboration with teachers, school administrators, and experts, with the intention of providing a clear and consistent framework to prepare students for college and the workforce.

Remillard commented that even though in the 50 states, the NCTM standards did influence what happened in each state's department of education, there was, over time, significant variation in the standards set by each state. This was a great concern of many in the United States. The Common Core Standards Initiative was a different effort marking the first time that there was a move toward a coherent set of standards for all of, or the majority of, states.

She noted the source of the initiative: a political organization (the NGA). Like the NCTM, this organization does not have policy-making authority. However, it is made up of state governors, so its members have relationships with decision-making bodies. They commissioned a set of standards that they would encourage their states to adopt.

The DoED was also interested in having shared standards. It cannot set standards, but it tied "Race to the Top," an important funding opportunity, to the adoption of shared standards. Many states wanted to apply for the Race to the Top grants, so they quickly adopted the CCSS.

Commercial publishers and noncommercial curriculum developers who want to develop instructional materials that will be purchased by local districts are finding that they need to make changes in their existing materials. Many of them are developing new materials. So the existence of the CCSS is influencing the development of textbooks and changes in textbooks. Local districts are starting to look to commercial markets to find resources to support their teachers. This is how commercial publishers are influencing the curriculum.

In this case, instead of NSF, a private foundation, the Gates Foundation, has played a very strong role in facilitating the CCSS implementation. It has provided some support for resource development, and earlier support for states to apply for Race to the Top funding.

Remillard summarized: In the United States, we are trying to create a common set of standards, but to do that, all the pieces that influence the system need to be aligned and come into play. Because the system comprises so many distributed pieces, there are many opportunities for slippage—for the pieces to be misaligned, or misinterpreted.

At the end of the presentation, Gardunia described the development of new assessments for the CCSS. In her view, because the United States values individualism and freedom of choice, instead of having one assessment, there are two groups who are in the process of writing assessments. The states are allowed to choose the assessments they will use. These were to be in place for the 2014–2015 school year.

It is still a work in progress, Gardunia concluded, and we are not sure how it will turn out.

Reflections on Korean and U.S. Presentations

After the Korean and U.S. sessions on mathematics curriculum, Hyman Bass of the University of Michigan and Gabriele Kaiser of the University of Hamburg presented their reflections.

Hyman Bass said he detected a certain amount of similarity between the United States and Korea for the development of standards in that both referred to a reduction in content. He did not think of it as much as reduction, but as more focus, and greater emphasis on processes or practices. Bass believes that a great deal of content can be covered, and developed in more depth and with more flexibility, if practices are well taught. Covering many small topics with a lot of breadth is not necessarily a way of acquiring more mathematical knowledge.

He offered a personal understanding about the situation in the United States and the Common Core. The fundamental contrast between the two countries is that Korea has a ministry of education staffed by mathematics educators, mathematicians, and others who are relatively stable in their positions. They do not change with each election, and they are professionals who devote themselves to educational problems.

As noted in the U.S. presentation, the United States has a tradition of local control in education that dates to its founding. When the economy was highly regional, that local control made a certain amount of sense. The mandate behind educational improvement in the United States, as in Korea, is mainly economic competitiveness. Both economies are now, at the very least, national, rather than regional. The United States is extremely mobile; students and families move across different regions. So to address educational improvement and reform in localized, uncoordinated ways is totally dysfunctional. It would be like building a national railway system with different gauge tracks in each state so that each train has to stop at the border, unload its cargo, and put it on a different kind of train. It makes no sense. This kind of dysfunctionality is witnessed in the educational system.

Why has the Common Core not usurped local control? The answer is that it is sponsored by the NGA and by the Council of Chief State School Officers. Each of the chief state school officers is like the minister of education for his or her state.

Why would the governors concede this authority and let the standards of their states be developed together with those of other states? If they make their own standards, they have to invest to create those standards, then they have to create assessments for the standards in each of their states—a very expensive proposition. The states already had to collaborate to develop regional assessments because it was too expensive to develop individual state assessments. So they were beginning to suffer the costs of this kind of fragmentation and eventually became convinced that they had to create something that was national in scope.

Another consequence of fragmentation was there were two forces in American education that were national rather than regional: Textbook publishing and commercial testing. Both were driven not by educational policy, but by profit making. To be viable, textbook publishers had to be able to sell their books to each state. Textbooks had to treat the union of all the state curricula, and consequently were very thick books that children had to carry to school or use only at school. One achievement of the Common Core will be to reduce the size and cost of those books. Another will be to make possible a coherent system of teacher education.

In closing, Bass made one last point. There have been many reforms in education in the United States. All have more or less failed, in his opinion. The primary reason for this failure was that in each case the reforms put new demands on teachers. It was assumed that if new standards and new textbooks were written, they could be given to teachers and teachers could be told, "Go do this." If an airline adopts a new aircraft, their pilots are taken offline for six months to a year to learn how to fly it. They are not given the aircraft and told, "Go fly this." In the United States, teachers are given mandates that involve new content, new teaching methods, new standards to meet, and new assessments, but little professional development or preparation.

Gabriele Kaiser gave her reflection from a European perspective, drawing on her knowledge of U.S. and Korean curricula. She noted that there is a long tradition of international comparative studies. For Europe, the English researcher, Sir Michael Sadler, was very important. He visited French, German, and Prussian schools, and wrote a very famous paper in 1900: "How Far Can We Learn Anything of Practical Value from the Study of Foreign Systems of Education?" He proposed to send future teachers at the end of their studies to German and French schools to study "methods of teaching and systems of education."

On the other hand, he said, "It is a great mistake to think, or imply, that one kind of education suits every nation alike." We have to be aware that we can learn from each other, but only to a certain extent.

Kaiser noted that there are significant distinctions between Korea and the United States in curricula, school practice, and teacher education. In considering Korean education, she relied heavily on a special issue of *ZDM* on "The Balance between Foundation and Creativity—Features of Korean Mathematics Education."

How should continental European traditions be classified with respect to these two countries? (Note that "continental Europe" does not include Great Britain, because it has a very strong and different tradition.) Because content knowledge plays a strong role in continental Europe, Kaiser considers it to have a strong commonality with Korea. In continental Europe, subject-related approaches in mathematics education have priority. These approaches are known as subject-related didactics of mathematics and are associated with German researchers of the 1970s such as Heinz Griesel and Hans-Joachim Vollrath. Especially in Eastern European countries, with their tradition of strong links to the German school system, there was a strong relation to mathematics as a subject and a closely related method of teaching.

In Korea, content knowledge has a similar importance. Where does this come from? What are the famous Korean researchers, what kind of explanations do they give us about where the origin of strong dominance of content knowledge comes from? One common explanation is, of course, Confucian Heritage Culture, which sees a teacher as an expert, as a scholar-teacher.[3] This kind of education is tied closely to content.

Relationships between U.S. or Western approaches to those in continental Europe are not entirely clear. Progressive education with child-centered approaches dominates in Scandinavian countries, and especially in North and South America, and countries shaped by American influences, for example, the Philippines and Singapore (Nebres, 2006). Individualism is important in this kind of teaching and the "acting subject" is especially prominent. Content is more in the background, although not totally in the background. Anglo-American countries have a tradition of pragmatism (Kaiser, 1999). When it comes to mathematics teaching, argumentation and proof is less important than in European traditions. Mathematical structure is less important in syllabi and daily teaching.

[3] In later discussion, Kaiser noted that an article in *ZDM* (Wong, Wong, and Wong, 2012) examines the question of how different kinds of philosophical traditions such as Confucianism, Buddhism, and Daoism might influence mathematics education. This article concludes "it is over-simplistic to draw causal relationships between schools of thought and social phenomena."

Didactics is a common core of continental European tradition, and there is a broader notion of "didactical knowledge" (pedagogical content knowledge) in continental European countries than in Anglo-Saxon and English-speaking countries. Compared to English-speaking countries, this broader, theory-guided knowledge has a very close connection to the subject of mathematics.

Generally speaking, didactics is a constitutive element of continental European educational approaches (Pepin, 1999). In his ICME talk, Bernard Hodgson mentioned three features of didactics described by Winsløw (2007):

1. Its epistemological character, that is, the central focus on specific knowledge in a different form, and the conditions for enabling pupils to acquire it.
2. Its relation to the background science (in this case, mathematics as a science).
3. Its actual and potential roles in relation to teaching and teachers (in particular, didactics as a design science and as a part of teacher education and the professional knowledge base).

These kinds of aspects are very apparent and important when it comes to didactics and its traditions.

These educational orientations influence curricula. In European curricula, the very strong normative orientation influences their structure: There is a long preamble with normatively based goals of mathematics education. There is a strong subject orientation. Forty years ago, there were three mathematical pillars in lower secondary education: number (deliberately called "arithmetic"), algebra, and geometry. Nowadays, these are enriched by probability, which is not called "data," but rather "probability and data." In the lower secondary level, algebra dominates. In the upper secondary level, calculus is compulsory in Germany. There is an intensive focus on argumentation and proof.

Continental European curricular traditions and Korean curricula have a remarkable relationship, in contrast to American traditions, amongst others. There are similar pillars in the new Korean curriculum for the lower secondary level (Lew et al., 2012) and continental European curricula (high relevance of functions and structural reflections) despite earlier treatment of other topics in Korea. There is a strong focus on mathematics content in traditional and innovative classrooms (e.g., Pang, 2012). There are intensive reflections on the learning of proofs and their gradual development (Kim and Ju, 2012). There is a strong focus on content knowledge in teacher education curricula (Kwon and Ju, 2012).

Kaiser concluded that these reflect commonalities in didactics of mathematics in Europe and Korea.

Implementation of the CCSS

Mary Kay Stein of the University of Pittsburgh and Stacie Kaichi-Imamura of the Hawaii State Department of Education addressed three questions related to implementation of the CCSS.

1. How do the CCSS differ from previous standards?
2. What issues are surfacing as the community begins to implement the CCSS?

3. How have textbooks been developed for the curriculum?

Mary Kay Stein began by noting that there are two broad kinds of standards in the United States. As noted earlier, the NCTM started the standards movement for mathematics and other subjects with the 1989 release of the *Curriculum and Evaluation Standards for School Mathematics*. This was followed by other NCTM documents: *Principles and Standards for School Mathematics* (2000); *Curriculum Focal Points for Prekindergarten through Grade 8 Mathematics: A Quest for Coherence* (2006); and *Focus in High School Mathematics: Reasoning and Sense Making* (2009). As mentioned earlier, the NCTM cannot mandate adoption of its standards. Nevertheless, Stein remarked, "I think everyone would agree that the NCTM standards have had very far-reaching influences on curriculum frameworks, professional development, classroom practices, and standards in most of the 50 states."

The state standards came into being in the 1990s, and then were enshrined into law with the No Child Left Behind (NCLB) Act of 2001. In that act, each state was required to establish standards for what students should know and be able to do, and to create assessments for measuring their attainment of those standards. Unlike the NCTM standards, state standards have the force of law. As the NCLB law stands now, all students must have been proficient in mathematics by the spring of 2014, or sanctions would have been applied. This testing and accountability dimension of NCLB is the object of controversy in the United States.

Differences from Previous Standards

Stein described four ways in which the CCSS differ from previous standards.

First, compared to NCTM and state standards, the CCSS represent fewer, clearer, and higher standards.[4] *Fewer* means less standards per grade level. This was in reaction to the state standards; many of them had become quite bloated as a result of the NCLB law. A commonly heard description of the U.S. curriculum was "a mile wide and an inch deep."[5] *Clearer* refers to the simpler and more precise language in which the CCSS are expressed. This is supposed to make it easier for teachers to communicate with each other, and with students, principals, and parents. *Higher* refers to more demanding content. Together these three features should, theoretically, set up teachers to spend *more* time helping students to understand *more* deeply a smaller set of goals for each grade level. Stein commented that she thought that her U.S. colleagues would agree that this is a huge change for U.S. education.

A second feature is that the CCSS aim to develop understanding over time. They do this by defining a set of K–12 pathways that students must be on to meet standards for college and career readiness by the end of high school. The pathways, in turn, are rooted in a set of learning progressions.[6] Learning progressions are defined as changes in the levels of student thinking as they move toward the goal of instruction in interaction with an ordered set of instructional tasks

---

[4] Gene Wilhoit's statement. Executive director of the Council of Chief State School Officers. July 1, 2009.

[5] This description is associated with a study of curricula released in 2000 by the Trends in International Mathematics and Science Study, and thus predates the NCLB.

[6] The CCSS state: "The development of these Standards began with research-based learning progressions detailing what is known today about how students' mathematical knowledge, skill, and understanding develop over time."

that are intended to encourage the students' thinking to grow from level to level toward the goal. The teacher's job is to keep students moving along these pathways. The idea is that the teacher will be helped by the two new assessments that are being developed. The hope is that these assessments will give not only a summative evaluation at the end of each year but also formative evaluation along the way, providing reports to teachers about whether their students are on the pathway or falling behind.[7]

A third distinguishing feature is the positioning of the standards for mathematical practice (SMPs). At the national presentation at ICME, Mike Shaughnessy, the past president of the NCTM, illustrated ways in which the SMPs grew out of previous NCTM standards. While agreeing that there is some continuity in the substance of the SMPs, Stein argued that there were also some discontinuities. She noted the relative isolation of the SMPs from the content. Compared to past standards, the practice standards are not integrated with the content standards, are less well specified, much less specific than the content standards, and the same across grade levels.

This limited guidance from the CCSS regarding how to integrate the standards for practice with the content standards is a concern. In its absence, guidance will be left to textbook publishers, Stein said. A related concern is that assessments will not give as much weight to practice standards as they do to the content standards. In that case, teachers will get the message that the practice standards are secondary to the content standards.

The fourth feature is the commonness of the Common Core. A related concern is political forces, which continue to challenge the Common Core on the basis of the U.S. Constitution. Another concern is the equality of opportunity that is supposed to emerge from these common standards. It will not materialize on its own. Instead, it will require the commitment of state departments of education, resources, and the professional development of teachers.

Emerging Issues

Stein briefly mentioned two issues related to implementation of the CCSS: (1) instructional programs that can support the transition to the new standards and (2) assessments of the new standards.

She reminded the audience that it was the summer of 2012, and there would be two full school years before the assessments would be administered in 2014–2015. How would schools handle this transition period, especially at the instructional program level? Ideally, along with the CCSS would be appropriate K–12 instructional programs.

Ideally, this coherent instructional system would spring into place full blown at the moment that it was needed. Instead, schools and districts are scrambling to not only build and enact instructional systems but also to assure that students get help if they are not ready for the grade-level knowledge and skills that will be taught.

[7] For formative features of the two assessments, see www.parcconline.org/non-summative-assessments and http://www.smarterbalanced.org/wordpress/wp-content/uploads/2012/02/Smarter-Balanced-Teachers.pdf.

The two assessment consortia have been laboring for the most part behind closed doors, yet much depends on these assessments. Past experience suggests that teachers will pay the most attention to the assessments once they are released, and the standards will recede into the background. Teachers will teach to the assessments, and the curriculum will narrow to *what is tested.*

The claim is that teaching to the test—with the Common Core test—will be a good thing, because it supposedly will represent higher, more demanding kind of content and more performance-based activity. Past experience suggests that without additional professional development, teachers will imitate the form of assessment items, but not their substance. A second concern is, Will the assessments yield student data that is useful for guiding teachers?

Other concerns were identified from a practitioner's viewpoint, that of **Stacie Kaichi-Imamura**, who is the mathematics resource teacher in Hawaii's Department of Education. Hawaii has been doing in-service work related to the CCSS for the past year and a half.

In the United States, elementary teachers are generalists without the content knowledge demanded by the CCSS. Historically, Hawaiian elementary teachers have relied heavily on textbooks. After the CCSS were released, teachers still used their old textbooks, which were not aligned. This poses a great problem for teachers that do not have the necessary mathematical content knowledge.

Professional development is another challenge. There are more than 13,000 teachers in the state of Hawaii, on several islands. Approximately 9,000 teach mathematics. Many Hawaiian teachers are not tech savvy and not used to having professional development material delivered by video, webinar, webcasting, or other online tools.

The CCSS are very different from the current state standards. Previously, Hawaiian standards changed only slightly from version to version. Although the NCTM process standards should have been a part of classroom teaching, teachers sometimes did not know what the process standards were because they appeared at the very end of the Hawaiian document. The new CCSS give the standards for mathematical practice in the front of the document.

The CCSS are very specific about content at each grade. Because of this, some teachers need help in realizing that their favorite topic or unit may not fall within the standards for their grade.

Textbooks

Stacie Kaichi-Imamura described the situation with textbooks. The CCSS were released in June 2010. Publishers have not yet created aligned textbooks. Many textbook companies do key word searches and match or add new material to their textbooks, without removing unnecessary material. And they stamp it "aligned to the Common Core."

Before the CCSS, textbook companies created textbooks for the larger markets such as Texas, Florida, California, and New York. Publishers also wrote for coverage. Because standards varied from state to state, smaller states, such as Hawaii, had to use these textbooks and make them work. These states had to add or delete material, and tweak lessons, depending on their standards.

From Kaichi-Imamura's perspective, the tables have been now been turned due to the commonness of the CCSS. Because Hawaii is one of 45 states adopting the CCSS, it is now in a position to ask publishers for what it wants. In particular, it is looking for alignment with the CCSS mathematical practices and problem solving. Hawaii was reviewing instructional materials at the time of the workshop and anticipated the following timetable:

> August 2012–January 2013. Initial review by outside vendor. Recommend all material that meets all requirements.
>
> February–March 2013. Second review by stakeholder groups. Will narrow previous list to six per grade band.
>
> April–May 2013. Final review by selected committees of mathematics teachers statewide, to recommend one to two textbooks per grade band.

In the meantime, in response to the lack of CCSS resources, states have been sharing information. Hawaii has been using the following resources:

- Illustrative Mathematics, http://www.IllustrativeMathematics.org.
- Progressions, http://ime.math.arizona.edu/progressions, which illustrate the progressions in the CCSS.
- Open Educational Resources, http://www.oercommons.org.

Comments

The presentations evoked a variety of comments on the CCSS and the challenges of implementing them. Some of those comments are given below.

Promising aspects of the CCSS

- The CCSS have provided opportunities to have more focused conversations about the process of teaching and learning in the classroom.
- In the United States, children are expected to understand the concepts behind the math, but there might be teachers who do not understand them. It is a challenge, but it is promising to see teachers growing with students.
- CCSS provides learning progressions. It is important to provide this information to teachers—why we teach this topic, why this concept is important—looking at it from the viewpoint of students' learning trajectories.
- A challenge is how to embed the eight mathematical practices, that is, the SMPs, which are new for teachers who learned math with another approach.

Professional development for the CCSS

- A productive direction for professional development is to document vivid images of instruction in regular content where these practices are integrated.
- In the United States, there are both bottom-up and top-down approaches for teacher development. With CCSS, districts and states have received funding to support this initiative through the Race to the Top grants. That money comes from the federal government to the states, then from the states into the school systems. That is how federal money has been directly affecting the top-down approach. Teachers identify their needs, and then they get together to help build the professional development opportunities and figure out what has to happen. That is at a district level, sort of top and bottom.

Challenges for implementation of the CCSS

- The challenge for the United States in developing textbooks and curriculum is that there is no equivalent of the Korea Institute for Curriculum and Evaluation (KICE). KICE is always monitoring and developing curricula. The United States does not have any similar organization playing this important role.
- The 45 states that have adopted the CCSS have the standards in common, but how the states and local districts support teachers is *not* necessarily going to be in common. That will be determined at the state level to some extent; to a very large extent it will be determined by the local school districts.
- There are about 15,000 education systems in the United States, all of which operate very differently, and all of which have to integrate with different layers of different systems.
- The U.S. education system is highly susceptible to political pressure, and because of that, often not very research-based. Professional development is one of the most difficult pieces of the U.S. system because it is least regulated. Even though there is a lot of professional development activity based on the CCSS occurring all over the country, its quality is highly variable.
- It is known that professional development is most effective when you prepare teachers for the curriculum that they will be teaching. That was impossible to do at scale in the United States because there were so many curricula. There are almost 2,000 different sites in the United States where teachers are prepared. A new Common Core Standards for teacher education and development is needed.

4
DEVELOPMENT OF INSTRUCTIONAL MATERIALS IN KOREA

A group consisting of Kyong Mi Choi of the University of Iowa, Kyeong-Hwa Lee of Seoul National University, Kyungmee Park of Hongik University, and Ji Won Son of the University of Tennessee at Knoxville collaborated on presentations given by Son and Choi that focused on three questions.

1. What is the role of learning trajectories in the development of curriculum?
2. How are curricular tasks designed and what criteria are used?
3. How do you assure procedural competency and conceptual understanding in the design of the curriculum?

Learning Trajectories in the Development of Curriculum

Ji Won Son presented the group's discussion of learning trajectories. She noted that since Martin Simon introduced the term "learning trajectory" in 1995, the idea of learning trajectories has gained attention in the United States and Korea as a way to focus research on learning in the service of curriculum, instruction, and assessment. In particular, this is a way to bring coherence between learning and curriculum. For example, the Common Core State Standards for Mathematics document states:

> The development of these standards began with research-based learning progressions [here used as a synonym for "learning trajectories"] detailing what is known today about how students' mathematical knowledge, skill, and understanding develop over time. (2010, p. 4)

In the group's opinion, no specific term for learning trajectories exists in Korean mathematics education, but the ideas that underlie the notion of learning trajectory are not new (Empson, 2011).

In 1995, Simon described learning trajectories as having three components:

1. A set of mathematical goals.
2. A clearly marked developmental path.
3. A coherent set of instructional tasks or activities.

Similar components occur in the Korean national curriculum, Son said. The curriculum articulates the goals for student learning based on the big ideas of mathematics and thoroughly describes them. The curriculum and textbooks include descriptions of students' thinking as they learn to achieve specific goals in mathematical domains within and across grade levels. Also, the curriculum presents developmental paths, describing a typical learning route children follow in developing understanding of skills related to a particular mathematical goal. In Korea, the developmental path is called *GaeNyumdo* ("concept map"). The curriculum specifies sets of instructional tasks or activities aligned with the mathematical goals and developmental path.

Son gave examples of each, beginning with developmental paths for number and operation at the elementary level, focusing on fraction multiplication. Fraction multiplication is a grade 5 topic. But before learning fraction multiplication, students need to learn addition, subtraction, and multiplication with whole numbers, as shown in Figure 4-1. Figure 4-1 describes a curricular path, but at the same time emphasizes horizontal connections among various topics, including number and geometry. Table 4-1 shows some associated tasks and goals.

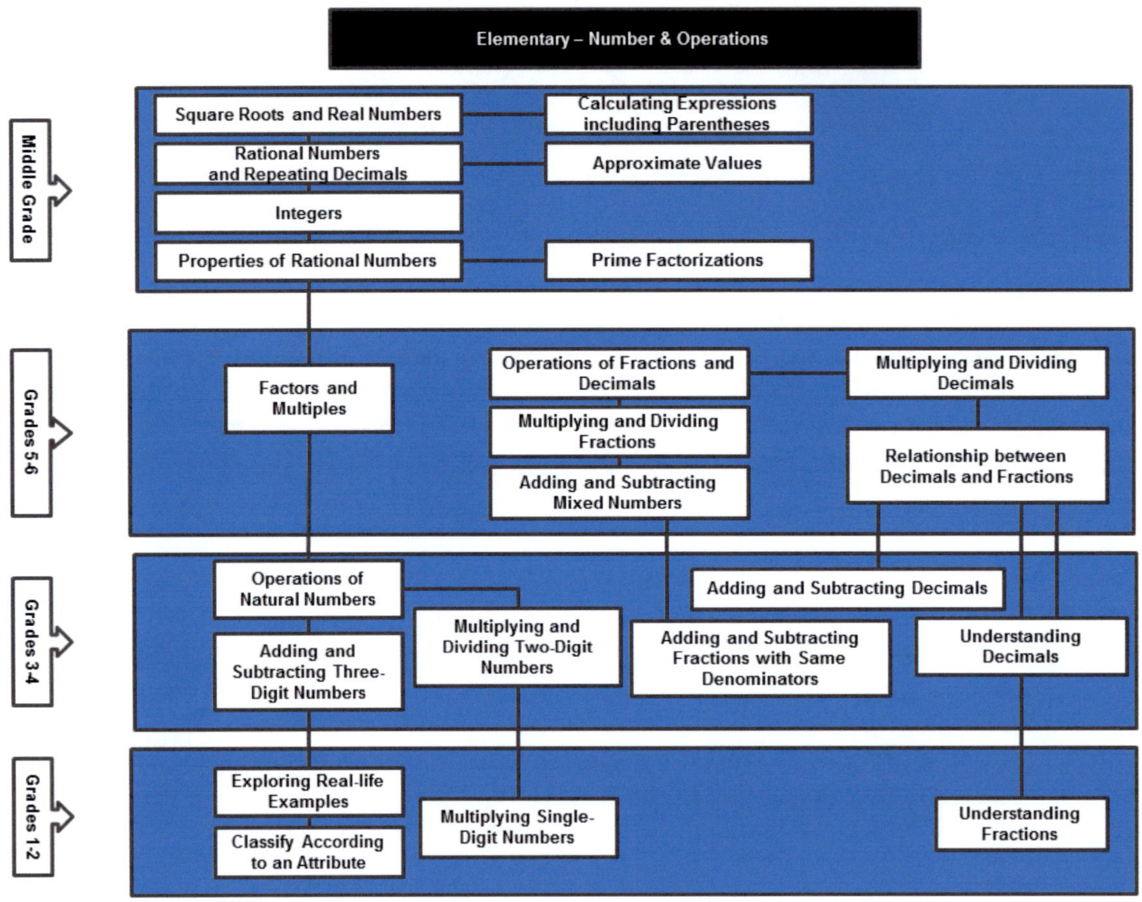

FIGURE 4-1 Curricular path showing connections among various topics. The developmental path is called *GaeNyumdo* ("concept map").
SOURCE: Kyong Mi Choi, Kyeong-Hwa Lee, Kyungmee Park, and Ji Won Son.

TABLE 4-1 Grade 5 Fraction Multiplication

Developmental Path (Mathematical Goals)	Instructional Tasks
1. To understand (whole number × proper fraction) and (whole number × mixed number), calculate it multiple ways.	There are five pizzas of which ⅜ of each pizza remain. How much pizza is there in total?
2. To understand (proper fraction × whole number) with manipulative; formulate algorithm and use it proficiently. To understand that if a multiplier is smaller than 1, then the product is smaller than the multiplicand.	12 m of wire was bought to make a wire sculptured animal with clay. If ¾ of the wire is used, how many m of the wire is used?
3. To understand (proper fraction × mixed number) with manipulative; calculate with two methods and compare two methods.	12-¾ m wire was bought to make a wire sculptured animal with clay. If ¾ of the wire is used, how many m of the wire is used?

Understanding fraction multiplication is broken down into three goals: understanding it as repeated addition, as an operator, and as finding a product of fractions. She provided a second example of *GaeNyumdo* at the secondary level, focusing on number and operation, and showing horizontal and vertical connections (see Figure 4-2), as well as goals and instructional tasks to teach irrational numbers (see Table 4-2).

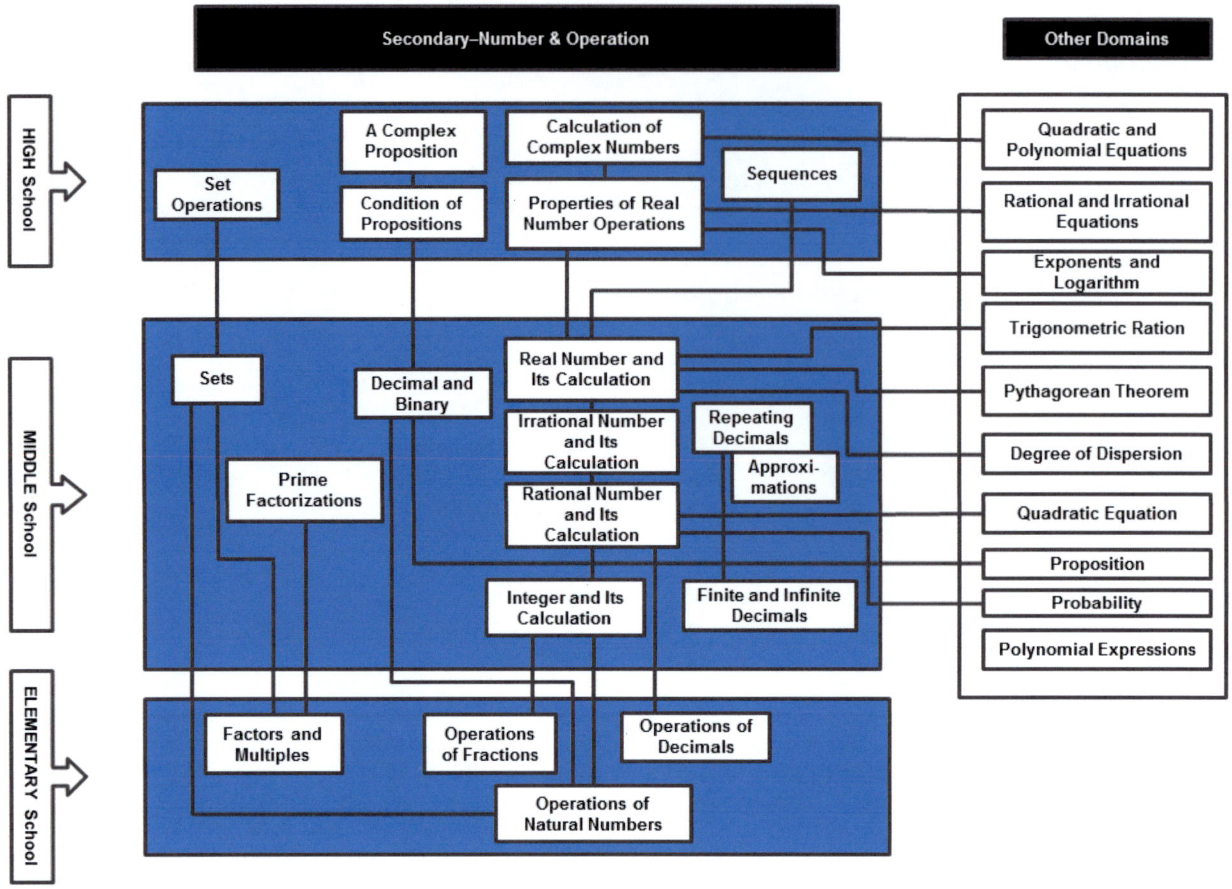

FIGURE 4-2 *GaeNyumdo* showing connections.
SOURCE: Kyong Mi Choi, Kyeong-Hwa Lee, Kyungmee Park, and Ji Won Son.

TABLE 4-2 Secondary: Irrational Numbers

Developmental Path (Mathematical Goals)	Instructional Tasks
1. To understand that there are numbers that are not rational; to know that they are called irrational.	A length of a side of a square is 1 cm. Can you express the length of its diagonal as a quotient of integers?
2. To understand and make rational approximations of irrational numbers in order to compare sizes of irrational numbers.	Locate each of the following on a number line: $\sqrt{3}, \pi, \frac{4}{3}, \sqrt[3]{2}, 0, \sqrt{16}, \frac{23}{5}, e$
3. To find sums and products of rational and irrational numbers.	For a rectangle whose length is $\sqrt{2}$ and width is $\sqrt{7}$, what would be the area of the rectangle? Would the product of two irrational numbers always be rational (or irrational)? Explain.

Son stated that her group thought that learning trajectories showed how students' conceptions developed—vertically and horizontally—by providing natural developmental building blocks from one level to the next. These help teachers and curriculum developers to assess, teach, and sequence activities. In particular, Korean mathematics textbooks provide *GaeNyumdo* at the end of the unit. These help teachers pose questions such as the following:

- What does this student understand?
- What could this student learn next and how could they learn it?

Dealing with such questions helps a teacher to consider learning trajectories.

Task Design

Kyong Mi Choi described features of instructional task design. Tasks are developed by teams of mathematicians, mathematics educators, and mathematics teachers to ensure mathematical correctness, and to consider how learners understand mathematics and possible misconceptions.

Tasks have five core elements: intuitive exploration, explanation, examples, practice, and extension. To promote *intuitive exploration*, tasks borrow concepts from everyday life, fill in a missing piece of reasoning, or use mathematical knowledge from earlier grades. For example, the Korean words for "territory" and "region" are the same. To familiarize students with the mathematical meaning of "region," a Korean traditional door invokes students' prior knowledge of a related meaning.

(After the presentation, **Mi-Kyung Ju** remarked that relating mathematical terminology to familiar words is important for equity because terminology is often borrowed from Chinese, and children from less educated families are less likely to be familiar with Chinese.)

The goal of *explanation* is to inform students so that they can figure out principles and meaning. Explanation includes defining mathematical terms and focusing on mathematical structures to be able to conceptualize, find algorithms, or contextualize. Core ideas and relevant background information are provided.

Example involves having students work through a typical example that reflects the core idea, justifying each step. Students reflect on mathematical processes, reasoning, and writing solutions correctly.

In speaking of *practice*, Choi quoted Confucius: “Is it not a pleasure after all to practice in due time what one has learned?” (*Analects of Confucius*, I.1). Practice allows students to make “new” knowledge their own by emulating and repeating what they have learned. Changes of mathematical structures when practicing enable students to reason mathematically and to make it a habit to apply relevant previous knowledge. Choi remarked that practice may be a factor in the superior mathematics performance shown by Korean students in international studies.

Extension gives students opportunities to explore different aspects that might have been overlooked or omitted. Students are encouraged to use multiple perspectives through problem solving, reasoning, and communication to connect mathematics and real-life activities. Choi noted that this focuses on objectification of mathematical content and process.

Procedural Competency and Conceptual Understanding

Ji Won Son described Korean approaches to assuring procedural fluency and conceptual understanding in the design of the curriculum. *Conceptual understanding* means comprehension of the *meaning* of mathematical concepts and operations. *Procedural fluency* refers to students' skill in carrying out procedures flexibly, accurately, and appropriately (Kilpatrick, Swafford, and Findell, 2001). In particular, the National Council of Teachers of Mathematics stresses that instruction should begin with a conceptual rationale for procedures, as the algorithms alone do not help students think about the operations and what they mean.

However, in teaching and learning of operations, Korean curriculum and instruction aims to develop conceptual understanding and procedural fluency simultaneously (Son, 2011, 2012) by use of the following four approaches:

1. Instructional tasks in each lesson are designed based on the three-step method:
   - i. Develop the meaning of the operation.
   - ii. Develop strategies for computing.
   - iii. Practice strategies and strategy selection.
2. Use a variety of representational models in a systematic way.

3. Encourage various types of computational strategies using distributive, associative, and commutative rules.

4. Use estimation frequently as basis for mathematical learning.

For example, when teaching fraction multiplication, the first emphasis is on repeated addition. Korean students solve problems using "area models" as sectors of disks. In Korea, "area models" may be disks and sectors of disks, while in the U.S., "area models" tend to be rectangles (perhaps composed of smaller rectangles).

Different representations are used in accordance with the three different meanings of fraction multiplication.

1. Repeated addition: area model.
2. Multiplication as operator: measurement and set model.
3. Taking a part of a part of a whole: area model.

To encourage use of various computational strategies, textbooks provide various methods for calculation, ask students to decide which are more efficient, and provide ways of making calculations easier (e.g., canceling before multiplying), and emphasize the use of the distributive property. Typically U.S. textbooks provide only one reduction method or none, Son stated, but Korean mathematics textbooks provide at least three reduction methods.

Use of estimation is frequent in order to help students understand concepts as well as to develop procedural fluency. Before asking the exact answer, textbooks ask students to estimate.[1] For example: Estimate how much this is: $\frac{1}{2} \times \frac{1}{3}$. Why do you think so?

Son ended the session with a summary. Korean curriculum and instruction provides opportunities to learn concepts and procedures simultaneously, rather than sequentially, by providing multiple computational strategies, using different types of representational models, making connections among ideas, and using number relations.

Comments

The discussion of developing instructional materials in Korea generated a variety of comments, many of which are given below.

Estimation

- Koreans found that when teaching estimation and fractions, they are trying to have students sense without calculating and develop the habit of thinking before calculating. Educators found that, with a little practice, children develop such a sense of estimation very fast, a lot faster than adults.

[1] The comments at the end of this chapter describe the different types of estimation and give some details about related mathematical goals.

Spiral curriculum

- Koreans have carefully implemented a spiral curriculum, starting in grade 2, with some preparatory work in grade 1, and it goes on until grade 5, instead of teaching fractions in the fifth grade like other countries. The Korean system indicates that a spiral structural approach functions better to teach mathematical content in a systematic, coherent, and consistent way.
- Both the United States and Korea follow spirals, but they may have different cycle lengths. Comparing the Korean with the Chinese curriculum, China deals with each topic in greater depth, more intensively. Therefore, the Chinese have fewer topics per grade than Koreans. In addition, the Chinese model is more linear than the Korean.

Teacher's manuals

- The teacher's manual in Korea does not seem to have the "Bible" status that textbooks do. Some teachers follow the teacher's manual with fidelity and some use it as a reference. There are many aspects of teaching that change spontaneously, even planned learning trajectories.
- Korean teachers are used to this lesson structure: introduce the idea of the lesson, develop ideas, and close the lesson. The textbook structure uses this format because teachers use it. Each textbook may take different approaches, but each lesson has a clear format: approach ideas, have students practice, then wrap up. Korean textbooks are very compact. More exercises, activity sheets, or tasks are provided by teacher's manuals, the teacher community, or curricular resources. In lessons, variation occurs in the selection of approaches, choice of tasks, and how the lesson is wrapped up, so the use of different textbooks does not matter to teachers.
- Korean teachers are required to do demonstration lessons where they show understanding of both horizontal and vertical trajectories. The beginning of the teacher's manual gives vertical trajectories for all six grades and prior knowledge at the beginnings of each unit. Teachers do the homework of reading the descriptions in the teacher's manual. Teachers have the belief that they can teach well only if they know what children learned previously, what they have to learn in this lesson, and what they will have to learn in the next lesson. Teachers do their homework sincerely and faithfully in preparing lessons, and analyze trajectories and apply those well in teaching.
- In the teacher's manual, Koreans explain why they put a particular task in a particular place, and give possible situations that could lead students to misunderstandings and the actions teachers can take in these situations.

Cognitive demand of Korean school mathematics

- Koreans made changes between the seventh curriculum and the 2007 revision, when they added differentiated workbooks according to student level. The textbook is aimed at average students, and the workbook is a supplement for upper- and lower-level students. In the 2009 revision, they decided not to use workbooks for the secondary level. Instead, they decided to raise demand in the textbook with higher-level questions.
- In Korea, the secondary level is worse than the elementary level in terms of cognitive demand and transparency in the teacher's manual.

5
TEACHERS, TEACHER PREPARATION, AND PROFESSIONAL DEVELOPMENT IN KOREA

The five presentations described in this chapter focused on teacher education from preparation to professional development. The main focus, teacher collaboration, was preceded by an overview of teacher preparation and hiring, and succeeded by an overview of main themes in Korean research on teacher practice.

The terms *professional development* (PD) and *professional learning community* (PLC) were used by workshop participants to describe activities in Korea and in the United States. In the workshop, the Koreans used three terms: PD, PLC, and the Korean word for "training." Both PLC and training were considered forms of PD. Training generally referred to attending required workshops and courses. PLC included nonmandatory PD such as teacher research groups. In Korea, these groups often receive government support and earn participants credit toward promotion.

Teacher Preparation and Hiring

Kyong Mi Choi of the University of Iowa, Mi-Kyung Ju of Hanyang University, and Oh Nam Kwon of Seoul National University created a brief overview of teacher preparation and the hiring process for public schools, drawing on Kwon and Ju's 2012 article on standards for professionalization of mathematics teachers in Korea.

The overview, which was presented by **Mi-Kyung Ju**, began by noting several findings from William Schmidt et al.'s 2007 report on middle school mathematics teacher preparation in six countries, including Korea and the United States.[1] Schmidt and his colleagues examined knowledge and course taking in mathematics, general pedagogy, and mathematical pedagogy. Future Korean middle school teachers scored well on content and pedagogical knowledge. Future U.S. middle school teachers studied only 43 percent of advanced mathematical topics compared to the 79–86 percent studied by their Taiwanese and Korean peers.

Ju described three stages in becoming a public school teacher in Korea. First, the prospective teacher, a high school graduate, must be admitted to a teacher preparation institute. Second, the future teacher is educated and graduates as a certified teacher. Third, the teacher must pass an employment test. Teachers are hired by provincial and metropolitan district offices of education. The number of teachers who pass the test is equal to the number of expected district vacancies for the coming academic year.

---

[1] The report cautions that "the U.S. sample is probably the least representative."

According to Ju, teacher education program entrants are among the top 10 percent of high school graduates. The passing rate for the employment test is one in 20.[2] So how can we ensure these high-quality entrants, who are highly educated in teacher preparation programs, will be excellent classroom teachers? Current discussion includes the question of whether Korean education has adequately nurtured prospective teachers with the knowledge and skills they need to be successful in the classroom.

Emphasis on classroom teaching has increased, said Ju. In 2006, the Ministry of Education, Science, and Technology (MEST) announced that teachers should be prepared more adequately in classroom teaching. In 2009, a new curriculum for teacher preparation programs was implemented, and the teacher employment process was altered to reflect the changes. To measure readiness, the second stage of screening was changed to include interviews and microteaching in order to measure educational disposition as a teacher, communication, and teaching skills. The first stage of screening is a written test. Part I consists of multiple-choice questions: 30 percent on general pedagogy and 60–70 percent on mathematics content and pedagogical content knowledge. Part II has four essay questions: 60–65 percent of the score is in mathematics, including proof writing, and 35–40 percent in mathematics education, for example, explaining how to teach a certain mathematical idea or how students approach a specific mathematical task. Ju concluded her presentation with examples of items from employment tests.

Policy Support for Teacher Collaboration

Rae Young Kim, a professor at Ewha Womans University, spoke about policies that affect teachers after they have been hired. As with hiring, requirements for promotion are uniform nationwide.[3] Before they start to teach, new (second rank) teachers must have 180 hours of training. After three years, before they are promoted to first rank, teachers must have further training. In addition to this training, teachers who participate in research groups earn one to two research credits toward promotion.

Studies have identified five factors involved in teacher collegiality (Ham, 2011; Hofstede, 2001; Marshall, 2004; Schmidt, Wang, and McKnight, 2005). At the national level, these are culture and curriculum policy; at the school level, climate and principal leadership; and at the teacher level, uncertainty management. Kim spoke about one of these factors: how Korean education policy provides opportunities and support for teacher collaboration.

Kim described two policies that support school-level collaboration: peer supervision and teachers' councils for curriculum. All teachers are required to participate in peer supervision, a form of planned demonstration lesson. Teachers must open their classrooms once or twice a year and obtain feedback from peers. The general format is as follows:

- Set up an annual plan.

---

[2] This is for the secondary teacher employment test (Kim, 2009, p. 50). The passing rate for elementary teachers is higher because only graduates from the 14 elementary teacher preparation colleges are eligible, which results in fewer test takers. Admission to elementary teacher preparation colleges is more competitive.

[3] There are several career paths. A teacher can become a master teacher, an assistant principal or principal, or a school commissioner or inspector in the research or administrative track (Kim, 2009).

- Prepare a lesson plan (individually or with other teachers).
- Observe and record the lesson.
- Discuss the lesson and receive feedback from peers.
- Revise the lesson plan.
- Open the class to teachers from other schools.

Each school has teachers' councils for instruction, formed by grade in elementary schools and by subject in secondary schools. These councils develop lesson plans and assessments cooperatively, and discuss classroom issues. They meet weekly and as needed to discuss issues arising while teaching in order to deal with them promptly. Kim noted that the extent and characteristics of teacher collaboration depend on their principal's leadership, the school climate, and the teachers' engagement.

Kim stated that Korean teachers have a strong desire to learn and to increase their capacity as teaching professionals. This desire is supported by MEST policies and by funding distributed by MEST and metropolitan, provincial, and county district offices of education.[4]

Interested teachers form school- or district-level groups and register them with their local district office. These groups research and develop new instructional methods and materials, which are disseminated by their district offices for other district teachers to share. Grants are available to curriculum research groups. These range from $5,000 for school-level groups to $10,000 for district-level research groups. Grants for other activities are available, but the emphasis is on curriculum and instruction.

District offices promote teacher collaboration by sponsoring research contests and teaching contests.[5] For example, in research contests for the improvement of instruction in Seoul, teams of four to five teachers compete by submitting innovative materials for enhancing teaching and learning or instructional methods.

There are many online archive sites run by government-funded institutes. Each site has a section on elementary or secondary school education, allowing information to be shared among teachers. Among these are the Korea Education and Research Information Service (KERIS). This houses the Research Information Sharing Service, which provides domestic and international databases for research; KERIS quality assurance, which controls e-learning quality; and EDUNET.[6]

---

[4] There are nine metropolitan and seven provincial offices of education for the 16 city and provincial districts. Under these are 182 county district offices of education. County districts are also known as local or regional districts. See http://eng.kedi.re.kr/khome/eng/network/domestic.do.

[5] Winners of teaching contests work as professional development leaders or as school consultants to improve teachers' classroom teaching. In Gyungi provincial district, winners are recognized with the title of "excellent teaching teacher." Teachers who win first prize three times receive the title of "treasured teacher." District offices, schools of education, and teacher organizations have teaching contests. In 2011, more than 10,000 teachers participated in a teaching contest held by the Korean teachers' union. Teaching contests often get media attention and reports in national television news.

[6] EDUNET is the largest education portal for teacher resources according to the site. It provides support for teachers, students, parents, and administrators. For K–12 students, it provides lectures, education software, and self-diagnostic tests in the five basic subjects of math, sciences, social studies, Korean, and English.

Teacher-Initiated Collaboration

Sooil Choi is the founder and a past president of the Korean Society of Teachers of Mathematics (KSTM), the largest mathematics teacher community in Korea. He discussed teacher-initiated collaboration, focusing on KSTM.

There are various teacher groups, communities, and organizations, and their numbers have been growing immensely. Some groups consist only of teachers; others also include university scholars. There are both online and offline communities. Some groups are formed within a school and some are not. Some are initiated by a government organization and some are initiated by teachers.

KSTM was founded in 1994 as Math Love and is still known by that name. Currently, about 10 percent of Korean mathematics teachers are members. It has 25 "common interest" subgroups, such as Classroom Observation and Analysis of Lessons, Elementary Mathematics, Development of Curriculum Material, International Studies, Mathematics and Culture, and Journal Editing. Choi participates in two subgroups: Development of Curriculum Material and Classroom Observation. Each meets once a week or has a weekly seminar. From the material they create, they provide professional development to teachers during summer and winter vacation. For the past 14 years, KSTM has held a math festival during the winter break.

In Choi's view, there are five ways that teachers can grow. The first three are individual efforts, such as (1) taking courses for advanced degrees, (2) attending teacher training workshops and lectures, or (3) reflecting alone on one's own lessons. But he noticed that individual effort does not help in classroom teaching or improving teacher quality as much as working together with other teachers. This is the motivation behind the creation of the KSTM subgroup Classroom Observation and Analysis of Lessons.

Lesson analysis is not easy to learn. Seven years ago, a format was created to help beginners learn key aspects of successful lesson analysis:

- Understanding the lesson plan.
- Preconference with the teacher, revision of the lesson plan.
- Classroom observation.
- Post-interviews with the teacher and students in the class.
- Analysis of the lesson.
- Feedback given to the teacher.
- Response from the teacher.
- Written summary of the results from the analysis.

After 2010, accompanied by the slogan "Everyone can do lesson analysis," the format was further simplified.

Choi said that members of the lesson analysis group developed a greater interest in students and deepened their identity as teacher professionals. There was a change in professional view and outlook. This was indicated as sharing beliefs in teaching, from mere observation of classroom

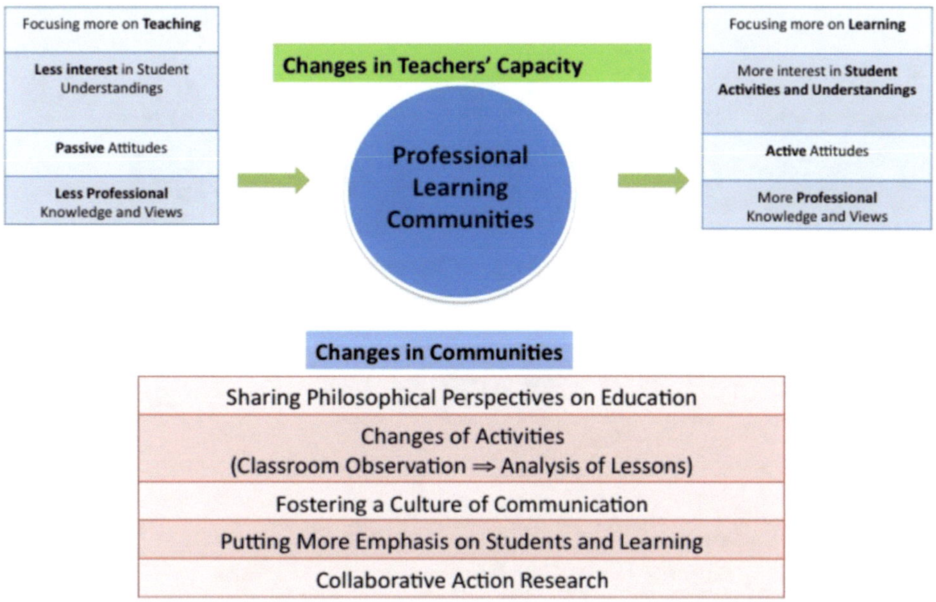

FIGURE 5-1 Changes in teachers' capacity as professionals.
SOURCE: Rae Young Kim.

teaching to critique and analysis, within a culture of collegial communities and group research, focusing more on students and learning, as illustrated in Figure 5-1.

Online Teacher Communities

Rae Young Kim briefly noted four of the many online communities created and managed by teachers.[7]

- Indischool, the main site for elementary teachers.
- GeoGebra Institute, a subgroup of KSTM for users of GeoGebra, a mathematics and science software.
- Kkulmat, an aid for study at home.
- A Group of Teachers for Gifted Students.

Lastly, she summarized policy- and teacher-initiated collaboration (see Figure 5-2). Korean teachers have a strong desire to learn and improve their teaching, content knowledge, pedagogical content knowledge, and instructional capacity. She thinks that is why they participate in many activities on- and offline. They collaborate with each other within and outside their schools. They have established communities by themselves and facilitate these communities. These are bottom-up movements, although teachers also have a lot of support from the government and their schools.

[7] Respectively, www.indischool.com; http://cafe.naver.com/geogebra; www.kkulmat.com; http://cafe.naver.com/xmot.

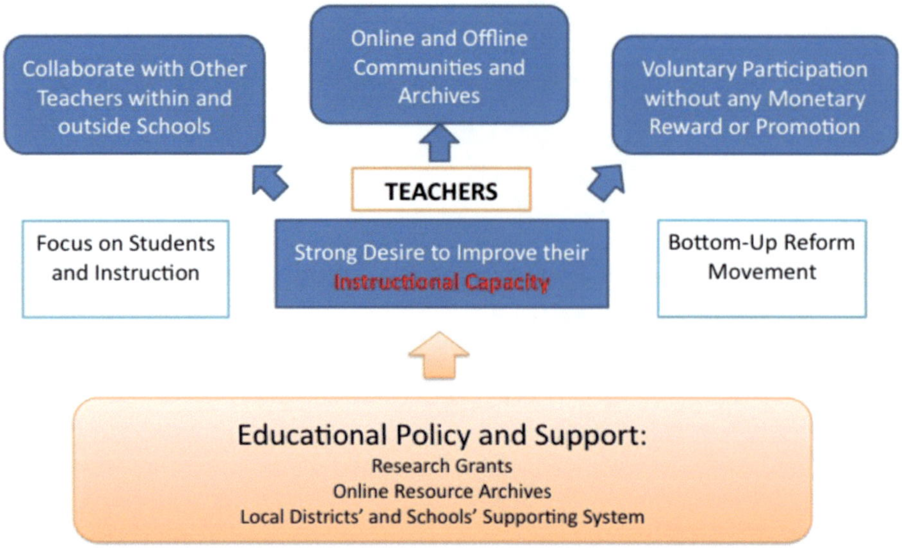

FIGURE 5-2 Teacher collaboration in Korea.
SOURCE: Rae Young Kim.

Kim ended by thanking the many graduate students who helped her create the presentation.

Research on Teaching Practice

Soo Jin Lee of Montclair State University presented an overview of Korean research on teaching practice. Lee began by noting that there used to be a widespread public perception that good teachers simply need to know a lot. But teaching is not a knowledge base; rather, it is an action, and teacher knowledge is only useful to the extent that it interacts productively with all the different variables in teaching (Ball and Cohen, 1999). Consistent with this observation, few studies have found correlations between teachers' knowledge and their students' achievement. This connection is mediated by practice, said Lee, and is a reason to investigate teaching practice.

Together with Gooyeon Kim of Sogang University, Na Young Kwon of Inha University, and Jae Hong Shin of the Korea National University of Education, Lee analyzed articles published between 2010 and 2012 in nine peer-reviewed Korean journals in mathematics education. Of the 150 articles examined, they focused on 40. They identified five emerging themes: (1) teacher questioning, (2) pedagogical tools, (3) pedagogical content knowledge, (4) reflection on teaching, and (5) social norms.

The studies on teacher questioning focused on communication and interaction among students and a teacher in classrooms. In the past, research was often based on the process–product tradition.[8] Recent research examines the types of communication used with questioning, and the virtues of pedagogical tools such as teaching strategies, instructional materials, and technology.

Studies of pedagogical content knowledge (PCK)[9] have examined relationships between teachers' PCK and their classroom instruction. The category of reflection on teaching includes research in which teachers reflected on their own teaching or that of colleagues whom they did or did not know.

A few studies investigated the culture of mathematics classrooms. Most were studies of social norms that used the emergent perspective of Cobb and Yackel (2004). One study used Lave and Wenger's perspective on situated cognition (Lave, 1988; Lave and Wenger, 1991) to investigate teacher change via participation in a community of practice in which elementary teachers shared their knowledge and experiences.

Lee concluded by emphasizing the need to investigate the teaching practice itself, rather than focus on specific components of teaching. She noted that such investigations require a theoretical framework for in-depth analysis of recurrent teachers' practices in classrooms. One example of such a framework is the teacher model mentioned in Alan Schoenfeld's presentation at the 12th International Congress on Mathematical Education.[10]

Comments

This segment of the workshop stimulated a variety of comments on teachers, teacher preparation, and professional development in Korea compared with the United States. Examples of the comments are given below.

Teacher career ladders, rotation at public schools, private schools

- In Korea there are second-class teachers and first-class teachers. To become a first-class teacher you have to do extra study and you have to actually pass an exam. In many districts in the United States, if you get a master's degree, then you are eligible for extra salary. The U.S. standards for the master's degree are quite diffuse. The Korean system is much more focused on strengthening subject matter than the U.S. system is.
- Korean teachers earn points or credits that go toward their promotions and the opportunity to be promoted throughout their careers. In the United States, teachers have opportunities to earn credits for recertification purposes; some will earn credits that will go toward their ability to become administrators, but that is not frequent.
- There are 16 provincial and metropolitan school districts in Korea. Teachers are hired and tenured in one of the 16 districts. Korean teachers in public schools rotate every five

---

[8] Research in this tradition investigates statistical relationships among quantifiable classroom processes (e.g., numbers of questions of various types asked by the teacher), and products such as students' scores on standardized tests. See, for example, Schoenfeld, 1994.

[9] Shulman (1986) defines PCK as "the particular form of content knowledge that embodies the aspects of content most germane to its teachability."

[10] "How We Think: A Theory of Human Decision-Making, with a Focus on Teaching." Available at http://www.icme12.org/upload/submission/1900_f.pdf.

years to keep the culture of the school fresh. That is, of course, within a region; teachers do not have to change their place of residence.[11]

- In Seoul, rotation is not a big problem, but in the south (where districts cover a large geographical area), moving from one area to another can be a problem. On the other hand, it guarantees equity. So every school has basically the same teacher quality because of rotation.
- In Korea, there are very few private elementary schools. There are more private schools at upper levels. The percentage of public schools in Korea is high: 59 percent at the high school level; 79 percent at the middle school level; and 98 percent at the elementary school level. Salaries of private school teachers are subsidized by the government.[12]

Professional development: PLCs and training

- In Korea, PLCs seem to be more bottom up, organized, and run by teachers voluntarily. The advantage of top-down [professional development] is that it gets more support from government, and so such PLCs are more likely to have a longer life. On the other hand, bottom up better reflects insiders' perspectives. The PLC has more freedom, but support might be a problem.
- There are many organizations in Korea, such as universities, teacher organizations, research organizations, government, and private industries that provide online teacher training. They develop programs and manage online systems for professional development. There are guidelines for online training courses. The programs and the number of credits associated with each workshop need to get government approval. The government subsidizes it so that it can control the quality. Teachers need to take courses with approved credits.
- In Korea, elementary teachers are prepared at 11 colleges, which specialize in professional development.
- In the United States, professional development is quite costly. Many teachers want to have professional development, but schools do not have the funds to offer it. There is a lot of professional development arranged in Korea that seems to be very cost-effective.

Korean teacher workload

- In elementary schools, Koreans teach 25 classes per week of 40 minutes each, five hours per day. Elementary teachers work from 8:40 a.m. until 4:40 p.m. Lower grade children go home right after lunch, which is 1 p.m. Upper grade students, in fifth and sixth grades, are dismissed at 2:30 p.m. Korean teachers spend three hours on out-of-classroom duties for lower elementary and two hours for upper elementary.
- Korean teachers prepare lessons in two ways: (a) preparing the next day's lesson at home and (b) at grade-level teacher meetings. They teach all subjects in elementary school, but during teacher education, they have a concentration subject. In the grade-level meetings, teachers allocate lesson preparation work and share prepared lessons based on their area of specialization.
- There is a nonprofit organization in Korea made with teacher initiative and support that provides material for all grades to any teacher, free of charge.

---

[11] For details, see the section on teacher rotation in Kim (2009).

[12] See Kim, 2009, pp. 31, 42–43.

- In Korean middle schools, teachers teach four classes per day of 45 minutes each, 20 hours per week. They teach the same course to five classes, so for one preparation they teach five times.
- There are not as many middle school websites with free resources as there are for elementary school in Korea. Usually, teachers do some research to find material online from various sites and develop their own lesson or modify one. Since there are many [secondary] textbooks, Korean teachers use several when preparing lessons.
- In high school, Koreans teach 15 to 16 classes per week of 50 minutes each. Grades 11 and 12 have elective mathematics courses. Regular school teaching is 15 to 16 hours per week. Besides teaching during school hours, educators have to teach after-school classes, so they also need to prepare for it. They spend lots of time researching lesson material. Twelfth-grade teachers prepare by solving an enormous number of different problems, many more than students would ever need to solve.[13]

Evaluation of teachers and schools

- In the United States, low-performing schools are often the schools that have to really stick by the textbook and are monitored more closely. Usually, flexibility in the United States is earned by high performance. Fewer funds are given from the ministry to Korean schools that perform poorly, but they often get help from university-based researchers that, in turn, are funded by MEST and foundations.
- Annual evaluation and measurement of student achievement is done by the Korea Institute for Curriculum and Evaluation (KICE).[14]
- Korea has a national uniform assessment of academic performance, which is administered every year. In July, all Korean students in grade 6 take a uniform national math ability test. The test is for study purposes and carries no penalty for teachers and schools. Its results are used when distributing government funding to the 16 districts, but do not affect school-level budgets. Districts with large improvement rates get allocated more. At the school level, the effect depends on the district; some districts allocated less to low-performing schools or give low points to all teachers in a low-growth school. Some districts do not use the results of this assessment.
- Academic performance of students in Korea is not reflected in school evaluations, but school activities are reflected in school evaluations. If a school is judged as providing a good quality education, it gets some extra funding. If an elementary school is determined to have low academic achievement, the government sends assistant teachers. Those students get extra help from assistant teachers after school, with textbooks developed for their level.
- Teachers in Korea are not penalized for low student performance. In fact, due to the education fever, they are trying to cool it down by inducing fun activities, such as math festivals, to increase student interest in the field.

---

[13] Generally, grade 12 lessons and a substantial part of grade 11 focus on preparation for the College Scholastic Ability Test.

[14] According to the KICE booklet, *Education in Korea*, "All sixth graders, third-year middle school students, and second-year high school students learning the National Common Basic Curriculum must participate in the NAEA [National Assessment of Educational Achievement]."

6
TEACHERS, PROFESSIONAL DEVELOPMENT, AND SPECIAL EDUCATION IN THE UNITED STATES

Karen King, director of research at the National Council of Teachers of Mathematics (NCTM), and Laura Ann Hulsebus, special education teacher and department chair at Alpenglow Elementary School in Anchorage, Alaska, spoke about the ways in which teachers' work is organized in the United States and how teachers work together to enact an intended curriculum.

King began with a national overview. This was followed by Hulsebus's description of her experiences as a special education teacher in very different school districts: one in rural Iowa and one in urban Alaska. After these presentations, the Korean participants commented, initiating a discussion with the U.S. participants.

National Overview

Karen King agreed with previous speakers that there is a tremendous variability in the United States. Within the 50 states, there are 14,000 public school districts and more than 90,000 schools. Each is a potential source of variability. This affects both the intended curriculum and how teachers work.

State textbook adoption was mentioned earlier. In a few states, the state selects textbooks and within those states, if a district wants to buy different textbooks, it must use its own money. If it wants to use the state's money to buy textbooks, it must select a textbook from the list provided by the state. Districts that buy their own textbooks add to the variation in the intended curriculum.

Curriculum development and guidance is local—generally at the district level. A district might adopt a state textbook and have the same standards as its state. However, it might make its own decisions about which lessons or units in textbooks to teach and the order in which these are taught. Policy guidance is local, usually at the district level.

Curriculum material is sometimes teacher created or what King called "teacher found." There are many sites where lessons and other curriculum material are posted on the Web, including the NCTM-sponsored site, Illuminations.

Thus, there may be differences in intended curriculum at the state, local, teacher, or student level.[1] Variability also occurs in school structures, teacher working conditions, and administrative expectations. How schools are organized and how teachers use their work time

[1] For further discussion of student-level variation, see the comments on special education at the end of the chapter.

can be very different from one school to the next, even within the same district or state. Therefore, it is very hard to paint a picture of what is typical in the United States. Research has described many factors that mediate progress from the intended to the enacted curriculum (cf., Stein et al., 2007). These include teacher knowledge (mathematical or pedagogical), time, teacher pedagogical skill, and local curriculum development and guidance.

Time varies. For example, some secondary schools have 45- or 50-minute class periods. Others have a "block scheduling," with 90-minute periods. This affects what happens to lessons, due to different amounts of time for exploration or discussion.

In contrast, if teachers offer children too much time to work on a problem, it can create classroom management problems. Another aspect of pedagogical skill is the setup of a task in the classroom. Others are management of classroom discourse, monitoring of student learning and engagement, and making adjustments to tasks along the way. All of these affect movement from the intended curriculum to the enacted curriculum.

Over the past 10 years, there has been a focus on teachers working together. A recent focus has been professional learning communities (PLCs): an infrastructure or a way of working together that results in continuous school improvement (Hord, 1997). These usually occur at the school level and vary from lesson study-like activities to common planning time. They may be entirely teacher directed and organized, or organized by the school. In terms of curriculum implementation, their focus may range from curriculum planning to the examination of student work to focus on the results of standardized and local assessments (Supovitz, 2002).

Cooperative activities such as these appear to be on the increase. In the 2007–2008 school year, 81.9 percent of public school teachers reported that there was a great deal of cooperative effort among staff, up from 77.5 percent in 1994–1995.[2]

PLCs may focus on larger school initiatives rather than on classroom instruction. Research suggests that in order to affect student achievement, the focus of a PLC needs to be instruction (Supovitz, 2002). However, other foci may affect teacher morale or interest.

Areas of focus related to classroom instruction include enhancing teachers' own mathematical knowledge for teaching; identifying and focusing on common student misunderstandings; and understanding the key mathematical ideas in the intended curriculum, specific pedagogical strategies for enacting that curriculum in the classroom, and assessment tasks and strategies to monitor student progress and inform instructional decision making.

There are barriers to the implementation of PLCs. These include insufficient time for the PLCs to meet. For example, the typical teacher in a high school may have one free planning period, which may not occur at the same time as his or her colleague's. Finding a time to meet together,

[2] U.S. Department of Education, National Center for Education Statistics, Schools and Staffing Survey: Public Teacher Questionnaire, selected years 1993–1994 through 2007–2008; Private Teacher Questionnaire, selected years 1993–1994 through 2007–2008; Charter Teacher Questionnaire, 1999–2000. See https://nces.ed.gov/surveys/sass/questionnaire.asp.

and, in some urban schools, a space, may be difficult. When teachers do meet, the meeting agenda may get diverted to other issues.

PLCs need administrative support and teacher leadership. There is considerable teacher and administrator mobility in the United States. Without broadly distributed leadership, movement of leaders may jeopardize the health of the PLC.

Lack of access to external resources may be a barrier. Some PLCs include mathematics educators and mathematicians from local universities, mathematics coaches, or others who provide support for teachers. This may be particularly important for PLCs with many very inexperienced teachers and not enough more experienced peers capable of supporting them. Another barrier may be lack of administrative leadership that prioritizes PLCs and monitors their progress. Prioritizing PLCs requires administrators to support them, to find ways to monitor the effectiveness of the PLC, and to seek out and support access to external resources. If the administration does not support these activities, then the PLC cannot go forward.

A Teacher's Perspective

Laura Anne Hulsebus described her 13 years of teaching. She reiterated the variability of the United States: Every single state is different, every single school district is different, and every single classroom is different. This variability is illustrated by her teaching career.

Her career began in rural Iowa, in a school of fewer than 300 students between prekindergarten and grade 12. This school was part of a consolidated district for five towns, which had kept two of five school buildings open. There was one special education teacher for each level—preschool and elementary, middle school, and high school. Before Hulsebus started, there were two special education teachers, one in the middle school building and one in the high school building. Elementary special education students would come to the middle school or high school for one class period as schedules permitted.

On her first day, Hulsebus was told that the school had never had a special education elementary teacher before and did not have a special education curriculum. She was given a room that looked like a closet with carpet on half the wall. The superintendent told her, "I'll be in my office if you need me."

There was very little professional development. Some of the teachers would coordinate, some would close their doors. The teachers had quarterly meetings to create the standards. All students were tested on the Iowa Tests of Basic Skills.

The school was literacy focused. At least two hours a day were spent on reading, and less than an hour on mathematics and science. Because Hulsebus was the only elementary teacher trained in assessment, she was required to do all the assessments for all the elementary students (about 100), and report the results to the state.

The special education teachers felt very isolated, said Hulsebus. They did not have anyone else to go to for support. The occupational therapist, physical therapist, school psychologist, and social workers were part of the area association and not part of the school district. To schedule meetings, people who lived in different towns needed to be contacted.

In the mornings, Hulsebus taught reading and mathematics in the same room to approximately 20 students in grades 3, 4, and 5. This included a gifted grade 5 student who was working at a high school level. Hulsebus had a 15-minute transfer to the next school, during which she ate lunch in her car. At the end of the day, she taught at a private Catholic school to which the public school provided services. She did not have time for planning.

After six years in Iowa, Hulsebus was ready to leave teaching because of the lack of support. As a last-ditch effort, she began teaching in Anchorage, Alaska.

The Anchorage school district is quite different. With just under 50,000 students, it is a much larger district than in Iowa. Depending on the case load, the number of special education teachers fluctuates. During 2011–2012, Hulsebus's school had two full-time and two part-time special education teachers. Two years earlier, there were four full-time special education teachers. The school district is large enough to have an autism classroom. Students who need additional services may attend specialized schools.

In contrast with her situation in Iowa, Hulsebus can choose among six different special education curricula for mathematics. Because there are specialized classrooms and schools, she is not required to teach all special education students.

In Anchorage, when a new curriculum is adopted, all teachers are given professional development. Teachers new to the district are required to attend professional development for each curriculum that they teach. Most of the time, teachers are paid to take professional development. There are addendums and graduate credit is provided for a reasonable price. For teachers like Hulsebus, this is a good incentive to keep taking additional professional development beyond what is required.

Depending on the building, some elementary teachers work together and some do not. In Hulsebus's building, about half of the teachers coordinate by grade level. There is block scheduling for language arts and mathematics, and a block time during which teachers can choose to meet.

The school psychologist, occupational therapist, physical therapist, and teacher consultant are all part of the Anchorage School District and are in her building at least once, sometimes twice, per week. There are monthly special education team meetings.

There are three grade 6 classrooms in her school. One is technology based with an interactive whiteboard and online lessons. One is traditional. The third is a social–emotional learning classroom that is "fully included." Half of the students are special education and two-thirds of the non–special-education students are at risk, Tier II, or nonproficient. In this classroom, Hulsebus team teaches with a general education teacher.

The grade 6 teachers all collaborate, trying to stay on the same mathematics lesson. (This is not typical for the school as a whole.) In placing students, the grade 5 teachers decide which teacher best fits the students' needs.

Comments

Comments following this discussion on teachers' professional development education expressed a wide range of ideas of what works and what does not in both the United States and Korea. Some of those comments are listed below.

Professional learning communities

- Some U.S. elementary schools allocate blocks of at least 45 minutes to an hour per grade level at least once a week to PLCs. Some schools are lucky enough to have it twice a week, and some get it every day.
- U.S. participants have been aware that U.S. classrooms are places of isolation. There is a history in the United States of some very strong professional learning communities taking place for 30 or 35 years, but they are in little pockets. Over time, the idea of PLCs and the idea that teachers may learn more from talking to one another about their practice than from sitting in a workshop have taken hold. Schools and school districts are starting to initiate PLCs in their different individual ways, providing opportunities for teachers to talk to one another. For example, they might start with a "problem of practice," where a teacher presents a problem of practice and others ask questions, discuss, and give feedback.
- American administrators sometimes have to force teachers out of their classrooms to make them participate in grade-level communication. Districts try to get at least one representative from every school to come and get professional development.

Professional development in the United States

- Providers of professional development have proliferated over the past 10 years or so. There are different teachers, often participating in different professional development and being held accountable for different actions, depending on the provider. That has been the subject of a lot of research in the United States.
- In the United States, there are three types of professional development: college and university level, district and state level, and national level. At the college and university level, teachers need to renew their licenses every five years, so they take courses—about six college credits.[3] These are not necessarily related to math and can be education courses. At the district and state levels, teachers do not need to pay, and they can take two workshops to renew their licenses. At the national level, professional development needs could be fulfilled by attending national conferences, such as NCTM. Sometimes teachers pay their own way, and sometimes they get school or school district support.

Special education

- The definition of special education is very similar in both countries.

---

[3] Renewal requirements vary by state. See Stillman and Blank, 2009, Table 16.

- Since 2009, special education in Korea has been gradually getting attention, and is required for [teacher] certification. Students with severe disabilities are placed in self-contained classrooms, and students with mild disabilities are placed in inclusion classes in regular schools. Students move from inclusion class to self-contained and vice versa.
- The United States has inclusion as well as resources for special education students.[4] For elementary education, Korea and the United States follow the same approach. However, for middle and high school, U.S. students are not excluded from regular education, but their parents choose to send their children to special schools for special education services.
- Special education can apply to any student who is not average; they could be above average (gifted students) or below average (students with various kinds of learning problems). In Korea and the United States, there are programs for gifted students, but in the United States, the programs for below average students are much larger and have a significant effect on the school system. These students very often require many more resources per individual to learn, so it is quite an expensive operation.
- Another effect is that it has increased the cost-per-pupil spending on education in the United States, but it has not increased the achievement level, because most of the extra money is going to low-achieving students. So special education has sort of created a new set of problems for United States education.
- What happens in the United States when the child receives a special education diagnosis is that she or he is given what is called an Individualized Education Program (IEP), which is where accommodations are listed. Once a child has an IEP, the educational institution must honor that. That is why it is called "accommodation." Any time there is an accommodation, it has to be listed in an IEP.
- Special education students generate more money for school districts than the regular education students. The IEP is a legally binding contract; anything that is written in an IEP stands up in a court of law. A lot of district money goes to lawsuits, if things are not followed on the IEP. So in the paperwork for a special-needs student, teachers feel like they are constantly proving that they are following the IEP to make sure that they are not sued.
- [Teacher certification for special education] differs from state to state in the United States. Some teachers might be certified just for "high-incidence special education," which usually means "learning disabled." In the United States, the Council for Exceptional Children is the organization primarily concerned with the education of teachers and with curriculum development for students with special needs.
- More and more students in America are being identified as special education and are in regular education classrooms. State by state, teacher programs are being required to include special education coursework in regular teacher education.
- Some U.S. districts have realized that special educators working in math classrooms may not have a mathematics degree or background.

---

[4] The Multi-Tiered System of Supports and the Response to Intervention are resources commonly used for special education students in the United States. See, e.g., http://www.districtadministration.com/article/multi-tier-system-supports and http://www.rtinetwork.org/learn/what/whatisrti.

7
OBSERVATIONS AND UNANSWERED QUESTIONS

In the final session of the workshop, U.S. and Korean participants presented what they had learned and questions that remained. Each presentation discussed the organization of the education system, curriculum and textbooks, and teachers and teaching practice, and ended with concluding remarks.

The U.S. Perspective

David Barnes of the National Council of Teachers of Mathematics (NCTM) presented remarks from the U.S. participants.

He noted that it was very helpful to see the structure of the Korean educational system, with its ministry and research arm. The United States does not have an organized research arm that studies education. One of the most important aspects of the Korean organization is that research is used to make educational decisions. In the United States, that is not always the case, even though relevant research may exist.

The U.S. participants also learned that Korea has regular cycles of curriculum analysis and change—sometimes large and sometimes incremental. In the United States, curriculum change happens when it is initiated by organizations such as NCTM. These changes occur at irregular intervals and with different processes.

Participants found differences in achievement by location very interesting. They noted the graph presented by Hee-Chan Lew showing that Programme for International Student Assessment (PISA) scores increase from rural to urban communities (see Chapter 2). That is not what occurs in the United States.[1]

Questions lingered concerning political forces in Korea and how they affect the educational system. In the United States, political forces seem to be having a growing influence on the educational system. Education is a political football and there are wild shifts. What happens in Korea? Questions also remained about the factors that influence the differences in achievement by locale.

Another collection of questions concerned university admissions tests, tutoring, and "education fever." What is the influence of university testing and its effect on teacher

---

[1] The 2009 PISA data show that large urban–rural score gaps did not occur for the United States, Belgium, Finland, Germany, Greece, Iceland, Ireland, Israel, the Netherlands, Poland, Sweden, and the United Kingdom. (See, e.g., *PISA 2009 Results: Executive Summary* [OECD, 2010].)

education and student learning? How does private tutoring influence the educational system? What about education fever? There was a comment about cooling it down in Korea. Perhaps it needs to be heated up a bit in the United States.

The inclusion of character building as part of the new curriculum, even the use of the term, was interesting for the U.S. participants. There were varying comments about the use of teachers in classrooms and testing of curriculum, but the U.S. participants were not quite sure what the process was. Other questions concerned the use of textbooks. Do Korean teachers have choices in selecting textbooks? How does the effort on individualized instruction and emphasis on creativity mesh with education fever? It seems as though the Koreans are trying to get the students involved and cultivate a love for mathematics. How do those fit?

Learning trajectories seem core to Korean curriculum. And there are very clear structures for lesson development. The five-step structure is consistent across the curriculum and across grade levels. To what extent are learning trajectories based on conceptual analysis of mathematics and to what extent on mathematics education research on student learning? Is there a melding of mathematical content development along with instruction? Do those contribute to learning trajectories? What is the basis for lesson design as part of the curriculum development movement and is there a mechanism for changing this structure?

Observations about teachers and teaching practices were that it was very clear that Korean teachers work together. This is embedded into the school; there is a location and time for teachers to work together. Professional development is expected, rather than required, that is what professional teachers do in Korea.

Questions concerned the small proportion of students that can become teachers and the employment test. If becoming a teacher is a very competitive process, is collaboration for preservice teachers a part of the program? Also, is there an expectation in the schools that could account for part of the urban–rural achievement difference?

Barnes ended by remarking that the education systems in both countries were in a state of change—moving toward each other. Both systems see the value of broader aspects of mathematics: creativity, problem solving, sense making, communication, proof, engaging students, and developing their love for and interest in mathematics.

The overall question that the U.S. participants had was about better understanding how the Korean system responds to struggling schools. How do Koreans address developing a positive attitude for students? How do they cope with problems of attitude?

The Korean Perspective

Rae Young Kim of Ewha Womans University spoke about the Koreans' reflections on the workshop.

She commented that the education systems in the United States and Korea have different characteristics, that is, uniformity vs. diversity, centralization vs. decentralization, fidelity vs. flexibility. The Korean curriculum seems to be changing rapidly. Action is taken promptly and quickly in alignment with the changes. On the other hand, the United States seems to go through curriculum change slowly and implementation takes a longer time. It might appear that the two countries have very different characteristics, but their directions seem to be converging toward the middle. The following table summarizes the key differences of the Korean and U.S. education systems, which were identified by the workshop participants.

TABLE 7-1 Key Observations of Korean and U.S. Education Systems

KOREA	UNITED STATES
Korea is a small country—like one very large U.S. state. The education system is very centralized, but has become less uniform regarding differentiated instruction and textbook adoption and development.	The United States has many different education systems, which operate very differently and have to coordinate at different levels.
There is one curriculum for all schools. The recent mathematics curriculum revision has fewer topics, treated in more depth.	Many states have adopted the Common Core State Standards (CCSS) for Mathematics, which have fewer topics, treated in more depth.
The government-affiliated Korea Institute for Curriculum and Evaluation (KICE) is a major textbook developer. KICE employs teachers, mathematicians, and its resident researchers to create textbooks and instructional materials.	Authors are currently creating revised versions of existing textbooks that they claim are in line with the CCSS. The content is reorganized, but its reduction is not evident. The United States has no equivalent of KICE to create textbooks and other instructional materials.
Many organizations provide online teacher training. Their programs and associated credits need to get government approval. For the recent curriculum revision, a government-affiliated institution (Korea Foundation for the Advancement of Science and Creativity) recruited teachers to design and carry out professional development.	The number of professional development providers has proliferated over the last 10 years. Many are "nonsystem actors" outside of the traditional bureaucracy of education. There is no uniformity in professional development opportunities for teachers. In some states, teachers help to deliver professional development.
Teachers treat the textbook like the Bible. They share offices with teachers of the same grade or subject and meet once a week at minimum. They have established their own communities. Each year, teachers are required to do demonstration lessons. There are many funds that promote teacher research. Research groups and individual teachers participate in competitions sponsored by district offices of education. Teachers rotate schools every five years.	Teachers pick and choose which parts of the textbook to use. They are isolated in their classrooms and sometimes act like independent contractors. Some very strong professional learning communities have existed for at least 30 years, but this is not widespread.

<table>
<tr><td>Students in grade 6, year 3 of middle school, and year 2 of high school learning the national curriculum must take the national test. Its scores have no effect on students. Instead, education fever is associated with college admissions tests.</td><td>Students are tested in grades 3 through 8, and at least once in high school. Consequences vary. In some states or districts, if the students do not pass the test, they do not go to the next grade. In others, students' test scores do not affect them.</td></tr>
<tr><td>Schools judged (according to test scores and school documents) as providing high-quality education get extra funding. Schools with many low-achieving students get other types of government support.</td><td>Student test performance can be part of teacher evaluation and evaluation of the principal and other school staff.</td></tr>
</table>

Unanswered questions concerned the development of the CCSS.[2] It appears that they were not developed by mathematicians or mathematics educators. Why were the CCSS developed by a third party? In the presentation about the CCSS, there was a mention of commonness. What is the meaning of common with respect to content, process, and pedagogy? What are the criteria for the commonness in the CCSS? Because the United States is a large and very diverse country, it must have been very difficult to come to an agreement. How did that process occur?

Ji Won Son of the University of Tennessee at Knoxville spoke about curricular tasks and classroom practice. There are many similarities between the two countries with respect to lessons, the structure of a lesson, and expectations for teachers or students.

Son noted that there is lot of research about learning trajectories in the United States, and asked how learning trajectories were reflected in the CCSS.

Even though there is much similarity between the two countries, since the cultures are so different, the Korean participants thought that the United States view of best practice must be different. How does the United States define best practice for teaching and for learning? Is there a rubric or structured system to evaluate teaching practice?

Videos of good teachers and teaching in the United States were shown at the workshop. What kind of roles do the teachers in the videos take in professional development? How do other teachers view them?

Jee Hyun Park of the Seoul Finance High School and Seoul National University presented remarks on teachers and teaching practice. In Korea, there are many forms of professional development: top down or teacher initiated. The professional learning communities (PLCs) seem quite different from these forms of professional development.

---

[2] The writings of William McCallum (CCSS lead writer) provide some answers about commonness, process, and learning trajectories. The CCSS for Mathematics for example, were commissioned by a third party, but developed by mathematicians, mathematics educators, teachers, and others. See *The Common Core State Standards in Mathematics* (ICME talk, http://math.arizona.edu/~wmc/Research/2012_07_12_ICME_McCallum.pdf) and *Testimony of William McCallum* (Wisconsin State Legislature, http://www.edexcellence.net/commentary/education-gadfly-daily/common-core-watch/william-mccallums-common-core-testimony-in).

While she heard about the U.S. situation, she thought: Why are PLCs so well developed in Korea? And why do Korean teachers attend professional development sessions so diligently?

In her opinion, it is not because teachers want to be better, but because of social pressure for teachers to have special expertise and the social expectation that teachers should do their jobs well. For teachers to renew themselves as good teachers, they must cultivate their professional knowledge—of mathematics, of classroom management, and of teaching—from their own internal initiative.

Jae Hong Shin of the Korea National University of Education was the last speaker. He said that he was given this opportunity because he stated that he learned more about Korea during the workshop. However, he thought there were many others who felt the same way.

He noted that mathematics education professors and teachers do not often talk as intensively as they did at the workshop. Moreover, they were able to talk with Koreans who work as professors in the United States. Shin himself came back to Korea two years before the workshop, after earning a degree from the University of Georgia. Much material about Korea was presented at the workshop from those with various kinds of different experience and expertise. That helped him to learn about Korea.

Comments

Final comments on observations of both the Korean and U.S. education systems are given below.

- Korean and U.S. systems are moving in the same direction and will eventually meet in the middle. The U.S. system is very fragmented and unfocused and is moving to the CCSS, while the Korean system is very focused (and somewhat rigid perhaps) and is moving toward a more flexible one, and opening toward more creative kinds of performances.
- Many classroom teachers in Korea participated in textbook revision. During the last revision, many participated in professional development and gave workshops about the new textbooks. Teachers' voices were heard and revisions were made that reflected comments from classroom teachers. Many teachers had a feeling of ownership, since they can influence policy and what is in textbooks. Korea encouraged [teacher] engagement and was connected with the effective and rapid implementation of the new curriculum and textbooks.

- In Korea [elementary] teachers are certified as generalists to teach all subject areas, but students preparing to be teachers major in one subject[3] (i.e., science education, math education, etc.).

[3] That one subject would be equal to about 22 credits in one concentration such as mathematics, science, Korean language, social studies, or music.

REFERENCES

Bae et al. 2011. *A Brief Understanding of Korean Educational Policy*. Seoul: Korean Educational Development Institute.

Ball, D. L., and D. Cohen. 1999. Developing practice, developing practitioners: Toward a practice-based theory of professional education. Pp. 3–22 in *Teaching as the Learning Profession: Handbook of Policy and Practice*, L. Darling-Hammond and G. Sykes, eds. San Francisco: Jossey-Bass Publishers.

Clements, D. 2007. Curriculum research: Toward a framework for "research-based curricula." *Journal for Research in Mathematics Education* 38(1):35–70.

Cobb, P., and E. Yackel. 1996. Constructivist, emergent, and sociocultural perspectives in the context of developmental research. *Educational Psychologist* 31:175–190. [Reprinted as pp. 208–226 in *Classics in Mathematics Education Research*, L. Darling-Hammond and G. Sykes, eds. 2004. Reston, VA: National Council of Teachers of Mathematics.]

Conference Board of the Mathematical Sciences. 2012. *The Mathematical Education of Teachers II*. Providence, RI, and Washington, DC: American Mathematical Society and Mathematical Association of America.

Council of Chief State School Officers and National Governors Association. 2010. *Common Core State Standards for Mathematics.* Washington, DC: Council of Chief State School Officers and the National Governors Association Center for Best Practices.

Empson, S. 2011. On the idea of learning trajectories: Promises and pitfalls. *The Mathematics Enthusiast* 8:571–596.

Gardner, D. P. 1983. *A Nation at Risk*. Washington, DC: The National Commission on Excellence in Education, US Department of Education.

Ham, S. 2011. *Collegiality as Uncertainty Management: Multilevel Contexts of Collaborative Teacher Interactions.* Ph.D. dissertation. Michigan State University.

Hofstede, G. 2001. *Culture's Consequences: Comparing Values, Behaviors, Institutions, and Organizations Across Nations* (2nd ed.). Thousand Oaks, CA: Sage Publications.

Honegger, S. D. 2010. *Revenues and Expenditures for Public Elementary and Secondary School Districts: School Year 2007–08 (Fiscal Year 2008)* (NCES 2010-323). Washington, DC: National Center for Education Statistics.

Hord, S. M. 1997. *Professional Learning Communities: Communities of Continuous Inquiry and Improvement*. Austin, TX: Southwest Educational Development Laboratory.

Huh, K. 2007. *School Curriculum in Korea*. Seoul: Korean Educational Development Institute.

Kaiser, G. 1999. Comparative studies on teaching mathematics in England and Germany. Pp. 140–150 in *International Comparisons in Mathematics Education*, G. Kaiser, E. Luna, and I. Huntley, eds. Philadelphia: Falmer Press.

Kilpatrick, J., J. Swafford, and B. Findell. 2001. *Adding It Up: Helping Children Learn Mathematics.* Washington, DC: National Academies Press.

Kim, D., and M. K. Ju. 2012. A changing trajectory of proof learning in the geometry inquiry classroom. *ZDM Mathematics Education* 44:149–160.

Kim, E. 2009. *Understanding Korean Educational Policy: Teacher Policy* (vol. 8). Seoul: Korean Educational Development Institute.

Kim, E., K. Kim, D. Kim, and E. Kim. 2008. *Improving School Leadership: Country Background Report for Korea*. Seoul: Korean Educational Development Institute.

Kim, R. Y. 2012. The quality of non-textual elements in mathematics textbooks: An exploratory comparison between South Korea and the United States. *ZDM Mathematics Education* 44:175–187.

Korea Institute for Curriculum and Evaluation. n.d. *Education in Korea*. Seoul: Author.

Kwon, O. N., and M. K. Ju. 2012. Standards for professionalization of mathematics teachers: Policy, curricula, and national teacher employment test in Korea. *ZDM Mathematics Education* 44:211–222.

Kwon, O. N., and S. J. Cho. 2012. Balance between foundations and creativity: Features of Korean mathematics education. ZDM Mathematics Education 44:105–108.

Lave, J. 1988. *Cognition in Practice: Mind, Mathematics and Culture in Everyday Life*. New York: Cambridge University Press.

Lave, J., and E. Wenger. 1991. *Situated Learning: Legitimate Peripheral Participation*. New York: Cambridge University Press.

Lew, H. C., W. Cho, H. Koh, and J. Paek. 2012. New challenges in the 2011 revised middle school curriculum of South Korea: Mathematics processes and mathematical attitude. *ZDM Mathematical Education* 44:109–119.

Lortie, D. 1975. *Schoolteacher: A Sociological Study*. Chicago: University of Chicago Press.

Ma, L. 1999. *Knowing and Teaching Elementary Mathematics: Teachers' Understanding in China and the United States*. Mahway, NJ: Lawrence Erlbaum Associates.

Ma, L. 2013. A critique of the structure of U.S. elementary school mathematics. *Notices of the American Mathematical Society* 60(10):1282–1296.

Marshall, M. L. 2004. *Examining School Climate: Defining Factors and Education Influences.* Georgia State University Center for School Safety, School Climate and Classroom Management.

Ministry of Education, Science, and Technology. 2011. *National Curriculum: Mathematics* (vol. 8) [in Korean]. Seoul: Author.

Ministry of Education, Science, and Technology. 2014. *2014 Report on School Violence: Survey and Analysis* [in Korean]. Seoul: Author.

National Academy of Sciences. 2010. *Preparing Teachers: Building Sound Evidence for Sound Policy.* Washington, DC: National Academies Press.

National Council of Teachers of Mathematics. 1989. *Curriculum and Evaluation Standards for School Mathematics*. Reston, VA: Author.

National Council of Teachers of Mathematics. 2000. *Principles and Standards for School Mathematics.* Reston, VA: Author.

National Council of Teachers of Mathematics. 2006. *Curriculum Focal Points for Prekindergarten through Grade 8 Mathematics: A Quest for Coherence.* Reston, VA: Author.

National Council of Teachers of Mathematics. 2009. *Focus in High School Mathematics: Reasoning and Sense Making*. Reston, VA: Author.

National Commission on Teaching and America's Future. 2003. *No Dream Denied: A Pledge to America's Children*. Washington, DC: Author.

National Research Council. 2011. *Incentives and Test-based Accountability in Education*. M. Hout and S. Elliot, eds., Board on Testing and Assessment. Washington, DC: National Academies Press.

Nebres, B. F. 2006. Philippine perspective on the ICMI comparative study. Pp. 277–284 in *Mathematics Education in Different Cultural Traditions: A Comparative Study of East Asia and the West*, F. K. S. Leung, F. Graf, and J. Lopez-Real, eds. New York: Springer.

Organisation for Economic Co-operation and Development. 2010. PISA 2009 Results: Executive Summary. Paris: Author.

Organisation for Economic Co-operation and Development. 2014. PISA 2012 Results in Focus. Paris: Author.

Pahlke, E., J. S. Hyde, and J. E. Mertz. 2013. The effects of single-sex compared with coeducational schooling on mathematics and science achievement: Data from Korea. *Journal of Educational Psychology* 105(2):444–452.

Pang, J. 2012. Changing teaching practices toward effective mathematics instruction in the Korean context: Characteristics and implications. *ZDM Mathematics Education* 44:137–148.

Pang, J. S. 2014. Changes to the Korean mathematics curriculum: Expectations and challenges. Pp. 261–277 in *Mathematics Curriculum in School Education*, Y. Li and G. Lappan, eds. Dordrecht: Springer.

Park, K. M. 1997. School mathematics curriculum in Korea. *Journal of the Korea Society of Mathematical Education Series D: Research in Mathematical Education* 1(1):43–59.

Pepin, B. 1999. The influence of national cultural traditions on pedagogy: Classroom practices in England, France and Germany. *Learners and Pedagogy* 237–252.

Sadler, M. E. 1900. *How Far Can We Learn Anything of Practical Value from the Study of Foreign Systems of Education?* Guildford, England: Surrey Advertiser Office.

Schmidt, W., et al. 2007. *The Preparation Gap: Teacher Education for Middle School Mathematics in Six Countries*. East Lansing, MI: Center for Research in Mathematics and Science Education, Michigan State University.

Schmidt, W., R. Houang, and L. Cogan. 2012. Preparing primary teachers in the United States: Balancing selection and preparation. *ZDM Mathematics Education* 44:265–276.

Schmidt, W., H. Wang, and C. McKnight. 2005. Curriculum coherence: An examination of US mathematics and science content standards from an international perspective. *Journal of Curriculum Studies* 37(5):525–559.

Schoenfeld, A. H. 1994. A discourse on methods. *Journal for Research in Mathematics Education* 25(6):697–710.

Schoenfeld, A. H. 1999. Looking toward the 21st century: Challenges of educational theory and practice. *Educational Researcher* 28(7):4–14.

Shulman, L. 1986. Those who understand: Knowledge growth in teaching. *Educational Researcher* 15(2):4–14.Simon, M. A. 1995. Reconstructing mathematics pedagogy from a constructivist perspective. *Journal for Research in Mathematics Education* 26:114–145.

Son, J. W. 2011. A global look at math instruction. *Teaching Children Mathematics* 17(6):360–370.

Son, J. W. 2012. A cross-national comparison of reform curricula in Korea and the US in terms of cognitive complexity: The case of fraction addition and subtraction. *ZDM Mathematics Education* 44:161–174.

Stein, M. K., and G. Kim. 2009. The role of mathematics curriculum materials in large-scale urban reform: An analysis of demands and opportunities for teacher learning. Pp. 37–55 in *Mathematics Teachers at Work: Connecting Curriculum Materials and Classroom Instruction*, J. T. Remillard, B. A. Herbel-Eisenmann, and G. M. Lloyd, eds. New York: Routledge.

Stein, M. K., J. Remillard, and M. Smith. 2007. How curriculum influences student learning. Pp. 319–369 in *Second Handbook of Research on Mathematics Teaching and Learning*, F. Lester, ed. Charlotte, NC: Information Age Publishing.

Stein, M. K., M. Smith, M. Henningsen, and E. Silver. 2000. *Implementing Standards-based Mathematics Instruction*. New York: Teachers College.

Stillman, L., and R. Blank. 2009. *Key State Education Policies on PK–12 Education: 2008.* Washington, DC: Council of Chief State School Officers.

Supovitz, J. A. 2002. Developing communities of instructional practice. *Teachers College Record* 104(8):1591–1626.

Winsløw, C. 2007. Didactics of mathematics: An epistemological approach to mathematics education. *Curriculum Journal* 18(4):523–536.

Wong, N.-Y., W. Y. Wong, and E. W. Y. Wong. 2012. What do the Chinese value in (mathematics) education? *ZDM Mathematics Education* 44:9–19.

Wu, H. H. 2011. Phoenix rising: Bringing the Common Core State Mathematics Standards to life. *American Educator* 35(3):3–13.

Appendix A

U.S.–South Korea Math Education Workshop Agenda

July 15–17, 2012
Seoul National University

Sunday, July 15 Overview (4:30–6:30)

15 minutes Welcome & Introductions

Korea – Oh Nam Kwon, Seoul National University

U.S. – Rick Scott, Chair, U.S. National Commission on Math Instruction

Myong-Hi Kim, U.S. National Commission on Math Instruction

Session 1 - The Math Curriculum in Korea and the U.S.: Key Issues

- An overview of the background and framework for the development of each country's math curriculum
- Issues and concerns in designing math curriculum

Korea – Hee-Chan Lew, Korea National University of Education
Man Goo Park, Seoul National University of Education
JeoungSuk Pang, Korea National University of Education
U.S. – Janine Remillard, University of Pennsylvania
Brenda Gardunia, Frank Church High School, Boise, Idaho

Background presentation from each country. (30 minutes per country)

Round Robin with all participants - What strikes you as similar and what is different from what happens in your country?

- Reflection from Mathematicians (15 minutes)

U.S. - Hyman Bass, University of Michigan, Past President of ICMI
Korea - Sung Je Cho, Seoul National University (Retired)
Germany - Gabriele Kaiser, University of Hamburg

Dinner 6:30–7:30

Monday, July 16 – Curriculum

Session 2 – New Curriculum and Textbooks (8:30–10:00)

- What are the main concerns (or main focus) in the development of the curriculum?
- What issues have been discussed or debated among curriculum developers, teachers, teacher educators, and scholars regarding the curriculum?
- How have textbooks been developed for the curriculum?
    - In particular, what are the main ideas used to support the learning of fractions?

South Korea (20 minutes): *Oh Nam Kwon, Seoul National University (SNU)*
Mi-Kyung Ju, Hanyang University; Rae Young Kim, Ewha Womans University
Jung Sook Park, Taenung High School; and Jee Hyun Park, SNU
Round Robin from the U.S. responding to "What do you see as challenging in what South Korea does for the U.S. and/or what is promising?" (20 minutes)

Open Discussion (50 minutes)

Break 10:00–10:30

Session 3 – New Curriculum and Textbooks (10:30–12:00)

- What is different about the CCSS from the standards or curriculum frameworks that have been in place?
- What issues have surfaced as the community begins to implement the CCSS?
- How have textbooks been developed for the curriculum?
    - In particular, what are the main ideas used to support the learning of fractions?

U.S. Presentation (20 minutes)
Mary Kay Stein, University of Pittsburgh
Stacie Kaichi-Imamura, Hawaii State Department of Education

Round Robin reaction from South Korea responding to "What do you see as challenging in what the U.S. does for South Korea and/or what is promising?" (20 minutes)

Open Discussion (50 minutes)

- Reflections on Sessions 2 and 3
  U.S. - Roger Howe, Yale University, ICMI Executive Committee
  Korea - Dohan Kim, Seowon, President of the Korean Mathematical Society

Lunch 12:15–1:00

Session 4 – Curricular Tasks (1:00–2:30)

- How are curricular tasks designed and what criteria are used?
- What is the role of learning trajectories in the development of curriculum?
- How do you assure procedural competency as well as conceptual understanding in the design of the curriculum?

Kyungmee Park, Hongik University; Kyeong-Hwa Lee, Seoul National University; Ji Won Son, University of Tennessee at Knoxville; Kyong Mi Choi, University of Iowa (20 minutes)

Round Robin reaction from U.S. responding to "What do you see as challenging in what South Korea does for the U.S. and/or what is promising?" (20 minutes)

Open Discussion (50 minutes)

Break 2:30–3:00

Session 5 – Classroom Practice (3:00–5:00)

Elementary:

Presiders: *David Barnes, National Council of Teachers of Mathematics*
Mi-Kyung Ju, Hanyang University

- 5-minute U.S. elementary video clip with subtitles.
- 2-minute writing what most surprised or interested you in the video.
- 5-minute Korea elementary video clip with subtitles.
- 2-minute writing what most surprised or interested you in the video.
- 30-minute Round Robin reaction from all participants.

Secondary:

Presiders: *Hyunyi Jung, Purdue University*
Rae Young Kim, Ewha Womans University

- 5-minute U.S. secondary video clip with subtitles.
- 2-minute writing what most surprised or interested you in the video.
- 5-minute South Korean secondary video clip with subtitles.
- 2-minute writing what most surprised or interested you in the video.

Round Robin reaction from all participants (30 minutes)

Open Discussion (45 minutes)

Dinner 6:00–7:00

Tuesday, July 17 – Teachers & Curriculum

Session 6 – Teaching Practices in the United States (8:30–10:00)

- How do teachers work together, if at all, in thinking about how to enact the intended curriculum? (i.e., cooperatively planned lessons, professional learning communities)
- What issues does this raise with respect to status, school support, equity across grade levels, schools, and districts?

U.S. Presentation (20 minutes)
Karen King, Director of Research, National Council of Teachers of Mathematics
Laura Ann Hulsebus, Alpenglow Elementary Anchorage, Alaska School District

Round Robin reaction from U.S. responding to "What do you see as challenging in what the U.S. does for South Korea and/or what is promising?" (20 minutes)

Open Discussion (50 minutes)

Break 10:00–10:30

Session 7 – Teaching Practices in South Korea (10:30–12:00)

- How do teachers work together, if at all, in thinking about how to enact the intended curriculum? (i.e., cooperatively planned lessons, professional learning communities)
- What issues does this raise with respect to status, school support, equity across grade levels, schools, and districts?

South Korea Presentation (20 minutes):
Gooyeon Kim, Sogang University; Na Young Kwon, Inha University
Jae Hong Shin, Korea National University of Education
Soo Jin Lee, Montclair State University; Sooil Choi, Past President of SMT
Dongwon Kim, KOFAC; Rae Young Kim, Ewha Womans University

Round Robin reaction from U.S. responding to "What do you see as challenging in what South Korea does for the U.S. and/or what is promising?" (20 minutes)

Open Discussion (50 minutes)

Lunch 12:00–1:00

Session 8 – Technology to Support Teachers and Students (1:00–2:30)

- How is technology being used to support the teaching and learning of mathematics?
- What is the role of research in how technology is being implemented?

U.S. Presentation (20 minutes)
Johnny Lott, University of Montana (Retired)
John Staley, Baltimore County Schools

Round Robin reaction from South Korea responding to "What do you see as challenging in what the U.S. does for South Korea and/or what is promising?" (20 minutes)

Open Discussion (50 minutes)

Break 2:30–3:00

Session 9 – Reflection and Sharing (3:00–4:30)

Each country will meet to discuss what they have learned and what questions remain to be clarified. (30 minutes)

U.S. Presentation on what has been learned and what questions remain to be answered (15 minutes)

South Korea Presentation on what has been learned and what questions remain to be answered (15 minutes)

Round Robin with each participant giving a closing comment (30 minutes)

Dinner 6:00–7:00

Appendix B
Biographies of Workshop Speakers

(at the time of the workshop)

Speakers from South Korea

Kyong Mi Choi is an assistant professor of mathematics education at the University of Iowa, Iowa City. She has master's and doctoral degrees in mathematics education from Teachers College, Columbia University, New York, and a bachelor of science degree in mathematics education from Seoul National University. She has taught high school mathematics in New Jersey. She served as a cochair of the 2012 National Mathematics and Science Competition, held by the Korean-American Science and Engineering Association. Her research interest includes mathematically gifted students and their learning experiences, mathematics curriculum analysis, and the Cognitive Diagnosis Modeling (CDM) approach on students' learning of mathematics.

Seoungyee Choi earned a bachelor of education degree in elementary mathematics education at Seoul National University and a master of education degree in gifted education (Further Mathematics) at Seoul National University of Education. She is pursuing a doctoral degree at Ewha Womans University and has taught elementary school mathematics for seven years at Joongdae Elementary School in Seoul. She has been in charge of the gifted class, operating the mathematics program for the gifted under the supervision of Gandong Education Support Agency since 2009, and is teaching mathematics to the students gifted in mathematics. In addition, she is the leading instructor for education trainees at Seoul National University of Education and one of the authors of the government-published textbooks. Her major concern is the development of elementary mathematics education programs for the gifted in mathematics and the interdisciplinary curriculum integration based on mathematics.

Sooil Choi is the director of research for the Mathematics Education Research Institute (MERI) and the founder of the Korean Society of Teachers of Mathematics (KSTM), the first association established by teachers for their own professional development purposes. He was the past president of the KSTM for eight years. He taught high school mathematics in public schools in Seoul for 27 years until he retired in 2011. He earned his Ph.D. in mathematics education from Seoul National University. He is currently a teaching consultant for secondary mathematics teachers. His interests include how to develop student-centered mathematics curriculum related to realistic contexts.

Mi-Kyung Ju is an associate professor of mathematics education at Hanyang University. She earned her Ph.D. in education at the University of California, Davis, in 2001. Her dissertation research was an ethnographic investigation of the social practice of professional mathematicians in an American university to describe the cultural production of academic mathematics. Her research interests include mathematics teacher training and preparation, and teaching and learning mathematics from a sociocultural perspective. Currently, in collaboration with mathematics teachers, she is involved in the development of an integrated curriculum and in the inquiry of teachers' professional development in the context of the curriculum development.

Hee-Chan Lew is a professor of mathematics education at Korea National University of Education since 1991. Previously, he was a researcher and a research fellow at the Korean Educational Development Institute. Lew has also held positions such as president of the Korea Society of Educational Studies in Mathematics, and member of the International Program Committee and cochair of the Local Organizing Committee for the 12th International Congress on Mathematical Education (ICME-12). He is also a member of the International Program Committee for ICME-13, which will be held in 2016 in Hamburg, Germany. Lew has directed projects on curriculum and textbook development funded by the Ministry of Education. Furthermore, he is the author and coauthor of Korean high school mathematics textbooks and more than 100 articles in various areas, including computer technology use in classrooms and professional development of mathematics teachers.

Kyeong-Hwa Lee is a professor of mathematics education at Seoul National University in Korea. She obtained her B.Ed., M.Ed., and Ph.D. from Seoul National University. Her research interests embrace mathematics curriculum and textbook analysis, mathematical reasoning and creativity development, gender issues in mathematics, and mathematics teachers' professional development. She has been involved in mathematics curriculum revision and mathematics textbook development in the primary and secondary levels since 1998. Also, she has been involved for many years in designing and implementing educational programs for mathematically talented students. She has studied gender differences in mathematics achievement and affective domain and has identified and nurtured female students who have potential in mathematics.

Gooyeon Kim is an associate professor of mathematics education at the Graduate School of Education at Sogang University in Seoul. She holds a Ph.D. in mathematics education from the University of Georgia. Her research focuses on mathematics teaching and learning. In particular, her research interests include how various types of teacher knowledge affect and are applied to mathematics teaching. Her research agenda also involves analysis of tasks in the curriculum materials and how teachers use and enact mathematics curriculum materials in mathematics classrooms.

Rae Young Kim is an assistant professor in the Department of Mathematics Education at Ewha Womans University. She was a mathematics teacher in public schools in the Seoul metropolitan area for seven years before she started her doctoral study. She holds her Ph.D. in mathematics education from Michigan State University, where she taught mathematics education courses for prospective teachers in the College of Education. Her academic interests include curriculum development, mathematics teacher education, and assessment for K–12 students in school settings. Her most recent research focuses on the development of interdisciplinary curriculum for teacher preparation programs and K–12 schools, as well as the development of analytic framework for nontextual elements in mathematics textbooks.

Na Young Kwon is an assistant professor of mathematics education at Inha University. Her research focuses on mathematics teacher learning and professional development. She is also interested in designing professional development programs for mathematics teachers and developing professional learning communities with teacher educators and mathematics teachers. Kwon earned her Ph.D. in mathematics education at the University of Georgia in 2007. She is working on improving the undergraduate program on mathematics teacher education in the Mathematics Education Department at Inha University.

Oh Nam Kwon is professor of mathematics education at Seoul National University. She received her Ph.D. in mathematics from Indiana University in 1993. She received her M.A. in mathematics from Seoul National University, her second M.A. in mathematics education from Indiana University, and her B.S. from Ewha Womans University. Her earlier professional appointments include assistant professor and associate professor of the Department of Mathematics Education at Ewha Womans University and visiting professor at Ohio State University and San Diego State University. She has been involved in more than 30 grants as principal investigator and collaborator. She is editor of the *Mathematics Education*, the *SNU Journal of Educational Research*, and the *Journal of the Korean School Mathematics Education Society*. She has served as a committee member for numerous international and Korean organizations of mathematics education, including the International Programme Committee of ICME-12. She is serving on the National Committee of the Korea Institute for Curriculum and Evaluation. She received the Best Teaching Award, Seoul National University, in 2009.

Jee Hyun Park has a bachelor of mathematics degree in secondary mathematics education and a master of education degree. She is a doctoral student in mathematics education at Seoul National University. She has taught middle and high school mathematics in Seoul for nearly 15 years. For the past three years, she has been on research supported by the Korea Foundation for the Advancement of Science and Creativity. She has been working on programs that develop secondary teachers' professionalism for nearly 10 years. She has been a teaching consultant of school consulting in the Seoul Metropolitan Office of Education for five years. Her professional interests are finding ways to increase mathematics literacy for secondary mathematics students and improving teacher preparation programs, especially teaching using technology.

Jung Sook Park has a master of education degree in mathematics education and has a doctoral degree in mathematics education. She had taught middle school mathematics for 17 years and has taught high school mathematics for 3 years. Her professional interests are finding ways to increase authentic learning experiences for high school mathematics students and improving teacher preparation programs to include stronger content knowledge for K–12 teachers, especially those that will be teaching middle and high school mathematics.

Kyungmee Park is a professor at Hongik University, teaching preservice teachers. She was a member of the Programme for International Student Assessment (PISA) Mathematics Expert Group from 1998 to 2004, and worked as a researcher at the Korea Institute for Curriculum and Evaluation. She has been involved in mathematics curriculum and textbook developments for the past 15 years. She writes columns about mathematics and education in major daily newspapers, and has contributed to the popularization of mathematics for the general public.

Man Goo Park is an associate professor at Seoul National University of Education. He is a coauthor of elementary mathematics textbooks as well as an author of book chapters on research in mathematics education. He is a coeditor of volumes 1 and 2 of *Mathematics Education in Korea* and has received three awards for his contributions to mathematics education.

JeoungSuk Pang is an associate professor of mathematics education at Korea National University of Education (KNUE). While her publications have been diverse, covering various aspects of elementary mathematics education, she is particularly interested in the analysis of mathematics classroom culture and professional development both for preservice and in-service teachers. Pang has been actively involved in developing the new mathematics curriculum and its concomitant textbook series. She received the Best Teaching Award at KNUE and is a member of the International Committee of the International Group for the Psychology of Mathematics Education (2008–2012).

Hyun Yong Shin is a professor of education at the Korea National University. He is the chair of the local organizing committee for the Twelfth International Congress on Mathematical Education (ICME-12). He has served as the president of the Korean Society of Mathematical Education. Professor Shin authored four university books for school teachers on set theory, number theory, linear algebra, and abstract algebra. He is also interested in the professional development of mathematics teachers. He received his Ph.D. in mathematics from the University of Alabama in 1987.

Jae Hong Shin is currently an assistant professor in the Mathematics Education Department at the Korea National University of Education. He earned his Ph.D. in mathematics education from the University of Georgia in 2010. His research interest is to investigate students' learning of mathematical concepts and to construct substantive explanatory models for their process of development. Specifically, his research focuses on identifying the mental operations that enable the construction of algebraic knowing, such as solving linear equations, combinatorial reasoning, and proportional reasoning. He is a member of the LOC Expert Advisory Committee of the ICME-12 in Korea.

Ji Won Son is an assistant professor of mathematics education at the University of Tennessee, Knoxville (UTK). She received her doctorate from Michigan State University with an emphasis in mathematics education, and before that she spent four years teaching elementary and middle school students in South Korea. Her areas of research include mathematics textbook analysis, elementary and secondary preservice teachers' knowledge development for teaching, in-service teachers' curriculum material use, and comparative study. She received a UTK Professional Development Award in 2009. She is currently serving on the program committee of the Association of Mathematics Teacher Educators (AMTE).

Speakers from the United States

David Barnes is the associate executive director for research, learning and development for the National Council of Teachers of Mathematics (NCTM). He joined the staff of NCTM in 2000 and at that time worked to bring the electronic versions of *Principles and Standards* to life. His involvement in NCTM has included expanding the online content of the council from journal articles to apps, to supporting the linking research and practice initiatives and the development of Research Briefs and Clips, to the development of the Navigations series and the creation of professional development institutes. Barnes holds a Ph.D. in math education from the University of Georgia and an M.S. in math from Illinois State University. His work has included the professional development of teachers at all levels, with a focus on the use of technology and problem solving, as well as work with teachers and schools on using innovative curriculum.

Hyman Bass is the Samuel Eilenberg Distinguished University Professor of Mathematics and Mathematics Education at the University of Michigan. He is a past president of the International Commission on Mathematical Instruction. His mathematical work is in various aspects of algebra, notably Algebraic K-theory. His work in mathematics education, largely in collaboration with Deborah Ball, concerns mathematical knowledge for teaching, the teaching of reasoning and proving, and ways to learn to see mathematical structure.

Brenda Gardunia has a bachelor of science degree in secondary mathematics education and a master of education degree in curriculum and instruction. She has taught high school mathematics in Boise, Idaho, for nearly 20 years. For the past two years, Gardunia has been on a leave of absence from the classroom and has been serving as an Albert Einstein Distinguished Educator Fellow at the National Science Foundation (NSF) in Arlington, Virginia, where, among other things, she has been working on programs that give research experiences to K–12 teachers and undergraduates. Her professional interests are finding ways to increase authentic learning experiences for high school mathematics students and improving teacher preparation programs to include stronger content knowledge for K–12 teachers, especially those that will be teaching elementary and middle school mathematics.

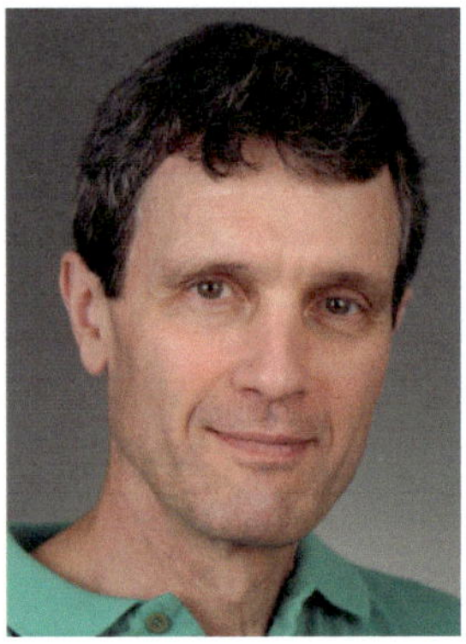

Roger Howe has been teaching and doing research in the Mathematics Department at Yale University for more than 35 years. His mathematical research investigates symmetry and its applications. He has held visiting positions at many universities and research institutes in the United States, Europe, and Asia. He is a member of the American Academy of Arts and Sciences and the National Academy of Sciences. Howe also devotes substantial time to issues of mathematics education. He has served on a multitude of committees, including those for several of the major reports on U.S. math education. He has served as a member and as chair of the Committee on Education of the American Mathematical Society. He served on the Steering Committee of the IAS/Park City Mathematics Institute, and has helped to organize a series of meetings devoted to increasing the contribution of mathematicians to mathematics education, especially to refining understanding of the mathematical issues in K–12 mathematics curricula. He is currently a member of the U.S. National Commission on Mathematics Instruction, and a member of the Executive Committee of the International Commission on Mathematics Instruction. In 2006, he received the Award for Distinguished Public Service from the American Mathematical Society.

Laura Ann Hulsebus is an elementary special education teacher in Eagle River, Alaska. She team-teaches a fully integrated sixth-grade classroom focusing on social-emotional learning and hands-on academics, and also a traditional resource program for kindergarten. Hulsebus is the department chair for special education at Alpenglow Elementary and math site base contact. She is also a special education teacher for summer school. Hulsebus has co-taught a graduate class for Anchorage school teachers on Writing Effective Assessments and participated in Big Wild Write Anchorage Teaching Teachers for training of the Alaska State Writing Consortium, which is part of the National Writing Project. Hulsebus's niche is teaching "twice-exceptional," Downs Syndrome and Asperger/autism students. She received the Presidential Award for Excellence in Math and Science Teaching for the State of Alaska in 2008.

Hyunyi Jung is a doctoral student in mathematics education and serves as the president of the Graduate Student Education Council (2012–2013) at Purdue University. She has been an instructor of supervised teaching in secondary math education and has taught math at the elementary, middle, and high school levels in the United States and in Korea. She received her M.S. degree in math education from Indiana University, Bloomington, while participating in the Secondary Transition to Teaching Program. Her experiences include being an observer for elementary teachers, a CCSSM Curriculum Alignment Tool Reviewer, and a coinstructor-translator for a math professional development program for Korean teachers. She currently participates in NSF-funded collaborative empirical research on teacher education, focusing on preparing to teach algebra and a design-based research project addressing mathematical modeling.

Stacie Kaichi-Imamura is the state mathematics resource teacher for the Hawaii Department of Education. She has been teaching for 17 years and has held multiple positions in the classroom in urban and suburban settings. Kaichi-Imamura has also been a school-level math coach and a district resource teacher. She holds a master's degree in curriculum and instruction from the University of Hawaii at Manoa. She is also a National Board Certified Teacher (Early Childhood Generalist) and a recipient of the 2004 Presidential Award for Excellence in Mathematics Teaching.

Karen D. King, Ph.D., is program director at the National Science Foundation in the Division of Research on Learning in Formal and Informal Settings in the Education and Human Resources Directorate. She most recently served as director of research for the National Council of Teachers of Mathematics. King's current research focuses on urban mathematics reform, the mathematics preparation of elementary and secondary teachers, and the policies of mathematics teacher professional development. King has served as associate editor of the *Journal for Research in Mathematics Education* and was a member of the RAND Mathematics Study Panel. She also serves on numerous committees focusing on research in mathematics education and teacher education with national organizations.

John Staley is the secondary coordinator of mathematics for Baltimore County Public Schools. As the coordinator, he works with a team of six talented individuals to support mathematics education in about 60 schools (28 middle, 27 high, several centers). He began his teaching career in Bucks County, Pennsylvania, just outside of Philadelphia, in 1987 at a juvenile correctional facility for males ranging in ages 12–18. He then relocated to Oxon Hill, Maryland, where he taught high school for a few years in Arlington, Virginia, and in 1993 relocated to Baltimore County. Over the past 24 years, he has taught middle school, high school, and college courses and now continues to teach secondary mathematics methods and courses at local universities.

Mary Kay Stein holds a joint appointment as professor of learning sciences and policy and senior scientist at the Learning Research and Development Center, both at the University of Pittsburgh in Pittsburgh, Pennsylvania. Her research focuses on mathematics teaching and learning in classrooms and the ways in which policy and organizational conditions shape teachers' instructional practice. Stein's most recent research examines how curricula can serve as a learning tool for teachers in large-scale improvement efforts.

Speakers and Participants Who Are Members of the U.S. National Commission on Mathematics Instruction

Gail Burrill is an academic specialist in the Division of Science and Mathematics Education at Michigan State University. Previously, she was a secondary teacher in suburban Milwaukee, Wisconsin and associate researcher at the University of Wisconsin-Madison. Ms. Burrill has over 25 years of experience teaching math education. In addition, she served as president of the National Council of Teachers of Mathematics (NCTM) from 1996 to 1998 and as director of the Mathematical Sciences Education Board at the National Academies from 1999 to 2000. Her honors include the Presidential Award for Excellence in Teaching Mathematics and the Wisconsin Distinguished Educator Award. She was elected fellow of the American Statistical Association and awarded an honorary doctorate degree from the Rose-Hulman Institute of Technology. Ms. Burrill received her MS in mathematics at Loyola University of Chicago in 1963.

Myong-Hi (Nina) Kim is an associate professor of mathematics at the State University of New York (SUNY), Old Westbury. At Old Westbury, she directed the mathematics teacher preparation program, and was instrumental in its accreditation by the National Council for Accreditation of Teacher Education and in having its newly created graduate program for teachers approved by the state of New York. She is a graduate of Yonsei University, Korea. She obtained a master's degree from SUNY Stony Brook and a Ph.D. from City University of New York under Michael Shub. She holds a New York State mathematics teacher's license. She has worked at the University of Southern California and Bell Communications Research, and has held visiting positions at IBM Watson Research Center and Pohang University of Science and Technology in Korea. Kim works in computational mathematics and received an NSF Professional Opportunity for Women in Research and Education grant. Kim was the principal investigator for a project funded by the International Korean Teaching Fellow Program of Korea's Ministry of Education, in which Korean teachers were placed in metropolitan New York schools that ranged from Stuyvesant High School to Francis Lewis High School. She is interested in improving college education, developing material for mathematics teachers, and cross-national interaction among mathematicians and mathematics educators from the United States, Korea, and developing countries.

Janine Remillard is an associate professor of mathematics education and outgoing chair of the Teaching, Learning, and Leadership division of the Graduate School of Education at the University of Pennsylvania (Penn-GSE). Her research focuses on mathematics teacher learning in urban classrooms and teachers' interactions with mathematics curriculum materials. She is primary faculty in Penn-GSE's urban teacher education program and is coeditor of the volume *Mathematics Teachers at Work: Connecting Curriculum Materials and Classroom Instruction.* She is the principal investigator of two NSF-funded studies: *Assessing Teachers' Pedagogical Design Capacity and Mathematics Curriculum Use* and *Learning About New Demands in Schools: Considering Algebra Policy Environments.* Remillard is vice chair of the U.S. National Commission on Mathematics Instruction.

Patrick (Rick) Scott is chair of the U.S. National Commission on Mathematics Instruction and vice president of the Interamerican Math Education Committee. He currently works for the Los Alamos National Laboratory Foundation as director of the Northern New Mexico Inquiry Science Education Consortium (ISEC). He retired in 2006 from New Mexico State University (NMSU) where he had worked as a professor of bilingual mathematics education in the College of Education. Upon retirement from NMSU he worked for the New Mexico Public Education Department where he organized a new Math and Science Bureau. In late 2008 he moved to the New Mexico Higher Education Department to be the director of P–20 Policy and Programs. From 1980 to 1996 he was a professor of mathematics education at the University of New Mexico.

Ana M. Ferreras is a senior program officer supporting the U.S. National Committees for mathematics instruction, crystallography, theoretical and applied mechanics, radio sciences, and physics. She joined the Academies staff in 2008. Ferreras holds a Ph.D. in industrial engineering (IE) from the University of Central Florida (UCF). She also holds an M.S. in engineering management from the Florida Institute of Technology and a B.S. in electrical engineering from UCF. During her doctoral research, she assisted the IE Department at UCF in reengineering the undergraduate curriculum by developing a national model, new programs, experiential laboratories, and research centers. Before joining the staff of the Board on International Scientific Organizations, she was a winter 2008 Christine Mirzayan Science and Technology Policy Graduate Fellow with the Center for Advancement of Scholarship on Engineering Education at the National Academy of Engineering.

Invited Guest

Gabriele Kaiser holds a master's degree as a teacher for mathematics and humanities for lower and upper secondary level, which she completed at the University of Kassel in 1978 with the first state degree. After having worked at school and completion of the second state degree, she worked as a scientific assistant at the Department of Mathematics at the University of Kassel, where she completed her doctorate in mathematics education (rer. nat.) in 1986 with a study on applications and modeling supervised by Werner Blum and Arnold Kirsch. Based on a grant for postdoctoral research by the German Research Foundation (DFG), she undertook her post-doctoral study in pedagogy on international comparative studies at the University of Kassel, which she completed in 1997. From 1996 to 1998 she held a guest professorship at the University of Potsdam. Since 1998, she has been full professor for mathematics education at the Faculty of Education, Psychology, and Human Movement of the University of Hamburg. Her areas of research include modeling and applications in schools, international comparative studies, gender and cultural aspects in mathematics education, and empirical research on teacher education. In October 2010 she became vice dean of the Faculty of Education, Psychology, and Human Movement, responsible for research, promotion of young researchers, and international cooperation. Since 2005, she has served as editor-in-chief of *ZDM—The International Journal on Mathematics Education* (formerly *Zentralblatt für Didaktik der Mathematik*), published by Springer. She is convenor of the 13th International Congress on Mathematics Education (ICME-13), which will take place in 2016 at the University of Hamburg.